THE COLORADO RIVER
THROUGH GRAND CANYON

Natural
History
and
Human
Change

THE
COLORADO RIVER
THROUGH
GRAND CANYON

Steven W. Carothers and Bryan T. Brown

The University of Arizona Press   Tucson

The University of Arizona Press
Copyright © 1991
The Arizona Board of Regents
All Rights Reserved
⊚ This book is printed on acid-free, archival-quality paper.
Manufactured in the United States of America.

96   95   94   93   92   91   6   5   4   3   2   1

Library of Congress Cataloging-in-Publication Data
Carothers, Steven W. (Steven Warren), 1943–
    The Colorado River Through Grand Canyon: Natural History and
Human Change / Steven W. Carothers and Bryan T. Brown.
       p.   cm.
    Includes bibliographical references and index.
    ISBN 0-8165-1131-4 (cl.). — ISBN 0-8165-1232-9 (pbk.)
    1. Stream ecology–Arizona — Glen Canyon Dam Region. 2. Stream
ecology — Arizona — Grand Canyon. 3. Stream ecology — Colorado
River (Colo.-Mexico) 4. Natural history — Arizona — Glen Canyon Dam
Region.
I. Brown, Bryan T. II. Title.
QH105.A65C38 1991
574.5′26323′097913 — dc20   90-46390
                                                                        CIP

British Library Cataloguing in Publication data are available.

To the scientists of the Glen Canyon Environmental Studies,
who went the extra mile

CONTENTS

ILLUSTRATIONS

FOREWORD

In midafternoon our river party pulled in at the head of Hance
Rapid, and we made our way downstream along the bank to
scout the water. Normally the river churns through Hance like a
freight train in a tunnel, reverberating through the chute and
then breaking into huge standing waves. But on this weekend,
Saturday, May 19, 1990, the water is trickling through like surf
draining off a rocky shore. There is no way that our wooden
dories can navigate this rock garden.

The alternative is to portage around the rapid by dragging the
boats up the left bank, across Hance Creek, over a couple of
huge sand dunes, through several acres of boulders and back
down to the water. No one is very enthusiastic, and a boatman
points out that there will be more low-water rock piles just
downstream at Sockdologer and Grapevine. Our leader, Martin
Litton, decides that we will camp and wait for high water.

We spread out on the beach. The evening is rich with the
sound of nesting birds, the lemon fragrance of flowering
mesquite trees, and Canyon walls glowing in the sunset. Perhaps
the Bureau of Reclamation did us a favor by turning the water
down.

The boatmen feel differently. As we unpack our rubber stuff-
bags, they are passing out pamphlets and earnestly explaining
the problem: the Bureau of Reclamation is holding water back at
Glen Canyon Dam, located just upstream from Grand Canyon,
in order to maximize electric power production at times of peak
demand. That means the river drops as much as ten feet each
night when air conditioners in Los Angeles and Phoenix go off.
And on weekends, when there is even less demand, the Bureau
shuts the gates still more, reducing the once mighty Colorado
River to about the size of a good irrigation canal.

When the Bureau of Reclamation opens the gates for peaking
power, a great surge of water heads downriver, stripping away
the sand and sediment along the banks, leaving barren mounds
of boulders and cobble. Sixty miles downstream an Anasazi ruin,
buried and preserved by river sediments for seven centuries, is

crumbling into the river. The Bureau, always ready with a new engineering idea, has suggested that it could stabilize the beaches by surrounding them with rock retaining walls. Or that it could build yet another dam, called a forebay, which would capture the surges from Glen Canyon and then release the water downstream in an even flow. All that a forebay would require is another hundred million dollars and drowning the last remaining stretch of Glen Canyon.

Among my fellow passengers is a writer for National Geographic who has spent weeks traveling the Canyon country. He reminds us that not all the changes have been detrimental. The river which now runs as ice cold and glass clear as an Arctic stream is full of trophy rainbow trout that never lived in the old muddy Colorado that we grew up with. In winter the spawning fish crowd together in the shallow tributaries like Nankoweap Creek, where flocks of bald eagles line the cliffs waiting their turn to feast on fresh fish.

And the majestic peregrine falcon, once at the edge of extinction, is now a common sight nesting in the cliffs and soaring above the river, in part because the smaller birds that it feeds upon have become more numerous in response to an expanded riparian zone of willows and tamarisk that provide nesting and feeding habitat. True as that may be, I remind him, national parks were not meant to be manmade zoological parks. Listen to the admonition of Theodore Roosevelt: "Leave it as it is. You cannot improve on it."

Now I am back in Phoenix enjoying that cheap air conditioning and reading and reflecting upon what Carothers and Brown have to say about this prolonged controversy. This book could not be more timely, for it summarizes a wealth of new information that is emerging from the scientific studies of the river corridor during recent years.

What emerges first from the studies is that the Colorado River is now an artificial, irrevocably transformed river, and there is no point in romanticizing about restoring it to its former, pre–Glen Canyon condition. Leave it to the monkey wrenchers to fantasize about blowing up the dam or re-creating the old muddy Colorado by scooping up silt from the bottom of Lake Powell and mixing it back into the water releases. The real issue is how best to regulate the releases to stop the destruction.

The authors suggest that much can be done. If daily and weekly river surging were reduced, it would be possible to restore sediment balance, prevent further bank stripping, stabilize the beaches, and protect native species. There is now

much evidence that the great uncontrolled floods of the early
1980s, caused by the Bureau's insistence on filling the reservoirs
too full, have been especially damaging to beaches and riparian
growth along the river. Particularly interesting is the suggestion
that proper storage guidelines and optimal timing of flood
releases could produce a better mix of native vegetation by
favoring the spread of willow and arrowweed, thereby
suppressing the continuing spread of tamarisk, an introduced
species.

In the long run, proper management of the Colorado River
will require institutional changes. The Grand Canyon river
corridor is being destroyed by the two federal agencies — the
Bureau of Reclamation and the Western Area Power
Administration — which have exclusive control over operation of
Glen Canyon. Each is obsessively committed to a single
objective to the exclusion of all others: Reclamation to store
every last drop of water and WAPA to generate every last dollar
of revenue as the water is released. Both agencies steadfastly
maintain that other public values and uses — boating, fishing,
threatened and endangered species, and the need to preserve
wilderness areas and national parks as unspoiled, pristine
remnants of the natural environment — are simply extraneous to
their mission.

After twenty years of court battles between environmental
groups and the two federal agencies, little has been resolved.
Only Congress can break the impasse, and at last there are some
signs that our leaders are awakening to the problem.
Congressional action to force proper environmental studies is a
beginning. However, the only lasting solution will be a
comprehensive management statute setting out multipurpose
criteria and environmental standards for the management of the
river and the operation of Glen Canyon Dam. Other federal
agencies, including the National Park Service and the Fish and
Wildlife Service, must have statutory authority to participate in
river management. State agencies responsible for game and fish
management should also be included. As the authors point out,
the successful experience with the federally created Northwest
Power Planning Council in the Columbia River basin may be a
good beginning point. Whatever the specifics, the time is at
hand to break up the fiefdom of the reclamation and power
agencies and to give the Colorado River over to true public
management.

BRUCE BABBITT

PREFACE

This is a book about change in the natural world of the
Colorado River through Grand Canyon. Changes are nothing
new for this ancient river, but the present rate of change due to
the recent actions of Glen Canyon Dam is unprecedented. Our
purpose is to tell the story of how the natural systems of the
river have adjusted to, or have persisted in spite of, the sweeping
environmental alterations brought about by the dam. We have
used dramatic examples from the river's interesting and well-
documented history to illustrate these alterations and to
speculate on those yet to come. The sum total of the recorded
history of the natural condition of the river, juxtaposed against
the stark post-dam changes brought to light by contemporary
scientific research, makes this story one of vital interest.

The beautiful and awe-inspiring portion of the Colorado
River between Lake Powell and Lake Mead provides a natural
laboratory for the grand experiment we wish to describe here.
This book explores what happens when a portion of a large
desert river is isolated between two dams — one upstream, one
downstream — and the water flowing through is carefully
controlled. What will be the nature of the new river that is a
product of this experiment? How will the water regime,
sediment budget, and food web of the new system differ from
those of pre-dam times? When will the new river reach
equilibrium, if ever? And most interesting of all, who or what
has made the river look and act the way it does today and the
way it will in the future?

Some of the answers to these questions may disappoint those
who expect the Río Colorado to represent an untouched
wilderness ecosystem. So be it. Yet the new Colorado River in
Grand Canyon is still a world-class ecological, recreational, and
spiritual resource, and to see the new river for what it really is
can never detract from that value. Most importantly, by
understanding these recent changes in the river, perhaps we will
be able to shape the river of the future for the better.

The idea for this book was first conceived by Steve Carothers
in the late 1960s, when he was a biologist working with the

Museum of Northern Arizona in Flagstaff. His repeated forays into the canyon to study the flora and fauna of the river suggested that the scope of the changes wrought by the dam was immense and that the story of those changes would have wide appeal. Bryan Brown began studying the river corridor in 1976, primarily its birds and their habitats, as an ecologist with the National Park Service. Together, we have had the combined good fortune to have completed more than one hundred raft trips through the canyon, for a total of more than six full years spent on the river in its study. And for better or worse, we have had the opportunity to observe firsthand the changes in the federal agencies and policies that have directly and indirectly influenced the river through that time.

Just as beginning boatmen have to learn how to "read the water" in order to navigate the river's rapids successfully, so we have had to learn to see the river's intricate systems to be able to understand the changes wrought by Glen Canyon Dam. This perspective was made possible by our many friends, colleagues, and adversaries, both on and off the river, to whom we are eternally grateful. Without their exploration and patient guidance, this book would not have been possible.

We would first like to thank those scientists involved in the Colorado River Research Program (1973–76) and the Glen Canyon Environmental Studies (beginning in 1983) for their dedication to studying and understanding the river. We have paraphrased many of their original research findings here. Though these contributions are too numerous to mention individually, we have cited their most important conclusions and trust that this will serve as full acknowledgment.

David L. Wegner, director of the Glen Canyon Environmental Studies, provided us with encouragement and support throughout the writing of this book. Nancy Brian and Mike Yard, also of the Glen Canyon Environmental Studies, were helpful in tracking down obscure information on the river and its biopolitical management.

Lawrence (Larry) E. Stevens, more than any other individual, has helped to shape our perception of how the river works. Dottie House gave us feedback and encouragement on each chapter of the book as it was drafted and helped substantially with the subsequent revisions.

Chapter 1 (water) was reviewed by Susan Werner Kieffer and David Wegner. Information on the flood of June 1983 was provided by Kim Crumbo. Chapter 2 (sand and sediment) was reviewed by Glenn Rink (who also reviewed the Introduction

and Chapter 1), Jack Schmidt, and Robert Webb. Kenneth
Hamblin generously shared unpublished information with us on
Pleistocene lava flows in the river corridor.

Chapters 3 and 4 (the aquatic ecosystem) were reviewed by
Bill Leibfried and Linn Montgomery; Dean Blinn, Gerald Cole,
and Larry Stevens also reviewed Chapter 3. The story of
trappers at Parashant Canyon in Chapter 4 was related to us by
Bob Euler, who also provided much of our information on
Stanton's Cave.

Chapter 5 (riparian vegetation) was reviewed by Art Phillips,
R. Roy Johnson, and Larry Stevens (all of whom provided
substantial unpublished information). Historical perspective on
the Nevills Expedition of 1938 (the Clover-Jotter Expedition)
was supplied by Nancy Nelson. Tom Moody inspired the design
for the plant-distribution chart in Chapter 5. Larry Stevens
provided much unpublished information on insect ecology
which was used in Chapter 6; he reviewed that chapter as well.

Ed Norton and Jim Ruch, of the Grand Canyon Trust, and
David Wegner, of the Glen Canyon Environmental Studies,
were instrumental in the development of the material presented
in Chapter 8 (biopolitics) and the epilogue. They and Rob
Elliott also reviewed early drafts of these sections. The entire
book benefited from reviews by the staff of Grand Canyon
National Park, including Kim Crumbo, Jerry Mitchell, and
Peter Rowlands.

The illustrations were created by Marilyn Hoff Stewart. Often
they were inspired by those prepared by the Colorado River
Research Program or Glen Canyon Environmental Studies.
Loren Haury, of the Scripps Oceanographic Institute, loaned
specimens of cladocerans for the preparation of illustrations in
Chapter 3. Slides of diatoms and *Gammarus* were obtained from
Bill Leibfried.

Photographs, or assistance in locating photographs, were
furnished by the Bureau of Reclamation (Upper and Lower
Colorado regions), Glen Canyon Environmental Studies,
Museum of Northern Arizona, National Park Service, National
Aeronautics and Space Administration, Special Collections
Library of Northern Arizona University, U.S. Forest Service,
and U.S. Geological Survey Photographic Archives in Denver.
Photographs were also provided by Buzz Belknap (Westwater
Books), Dugald Bremner, Christopher Brown, Harvey Butchart,
Michael Collier, Robert Euler, Susan Jones, Susan Kieffer, John
Running, Cecil Schwalbe, Curt Smith, Bill Suran, Ray Turner,
and Robert Webb.

The Colorado River Through Grand Canyon

Computer mosaic and enhancement of satellite photographs
by the National Aeronautics and Space Administration.

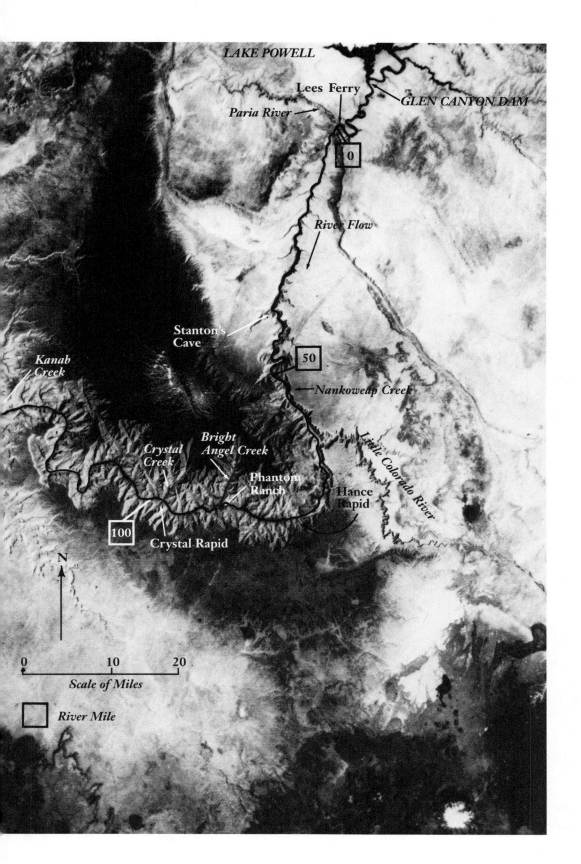

LAKE POWELL

GLEN CANYON DAM

Lees Ferry

Paria River

0

River Flow

Stanton's
Cave

50

Nankoweap Creek

Little Colorado River

*Kanab
Creek*

*Bright
Angel Creek*

*Crystal
Creek*

*Phantom
Ranch*

*Hance
Rapid*

100

Crystal Rapid

N

0   10   20
Scale of Miles

☐ *River Mile*

THE COLORADO RIVER
THROUGH GRAND CANYON

INTRODUCTION

Late in the hot afternoon of July 15, 1889, five exhausted and discouraged men in three wooden boats drifted down the Colorado River in Grand Canyon. They pulled ashore at the mouth of South Canyon, barely 30 miles downstream of Lees Ferry. Only the third group to attempt to traverse the Grand Canyon by boat, the expedition had proven to be nothing short of disastrous. Three members of the party had drowned in rapids during the seven days since leaving Lees Ferry.

Robert Brewster Stanton, an engineer, now led the ill-fated expedition. He had been hired by the original leader of the trip, Frank Mason Brown, president of the Denver, Colorado Canyon and Pacific Railroad Company and Stanton's good friend, to help explore and survey a route for a river-level railroad from Colorado to California. Brown, who drowned at the base of Soap Creek Rapid, was the first casualty. An inscription carved into the rock on river left at the mouth of Salt Water Wash, a half-mile below Soap Creek, still marks the site. Initially, Stanton had decided to carry on despite the loss, but the two additional drownings broke his spirit.

By the time the survivors reached South Canyon, Stanton knew they had been through enough. He gave the order to climb out of the tributary and head toward Kanab, Utah. Convinced that a river-level railroad was feasible, Stanton resolved to return and complete the survey. In anticipation of a second attempt, the party cached its equipment and remaining supplies just downstream of South Canyon in a large cavern known today as Stanton's Cave. The engineer did return a few months later and finished the survey, but for want of financial backing the railroad never was built.[1]

Today, Stanton's Cave is recognized for more than its role in this river runner's tale. The cave has proven to be a vault of the Grand Canyon's natural and human history, unparalleled in the wealth of information it stores. Stanton and his men probably were in a hurry to stash their gear and quit the canyon, for his journal's account of the day makes no mention of the cave's interior. Although it is easy to understand their lack of inquisitiveness, they overlooked some remarkable relics of the past. For scattered throughout the cave, some certainly only inches from the cached gear, were tiny split-twig figurines in the form of animals. These fetishes, and similar ones now known from other caves, represent the most common artifacts left in the canyon by the Pinto Basin people, a long-dead Native American culture. The artisans of the figurines are thought to be the first human inhabitants of the Grand Canyon. They were practicing their magic and weaving the ritualistic animal figurines found in Stanton's Cave almost four thousand years before Frank Mason Brown and Robert Brewster Stanton began their tragic trip.

The figurines would wait another forty-five years after Stanton's adventure before being discovered by later river runners and identified for what they were. The fetishes commanded attention, but the National Park Service soon realized that the cave held other items of prehistoric interest as well. Animal bones, centuries-old dung, and (even though the cave is 150 feet above the river) driftwood were scattered about the cavern. The driftwood was judged to be old, for recorded history revealed no evidence of a flood that could have allowed the Colorado River to rise to the level of the cave floor. Finally, in 1969 and 1970 an archaeologist named Robert Euler began the first scientific exploration of the cave.[2] Funded by the National Geographic Society, his team confirmed what was already suspected: Stanton's Cave contained artifacts that shed light on humans' earliest use of the Grand Canyon. In addition, pollen samples and fragments of bone and vegetation long buried in the cave-floor sediments constituted a time capsule of forty thousand years of natural history.

Careful examination of the extraordinary collection of debris unearthed during the excavation revealed ancient occupation by several extinct birds and mammals: the giant teratorn, a monstrous bird with a 12-foot wingspan; the ancestor of the California condor; the Harrington mountain goat; and many others. One of the most remarkable discoveries, though, came from an analysis of plant fragments found within the cave. With these relics, scientists were able to verify that the canyon's climate has changed significantly over the past millennia. At one time, a little over ten thousand years ago, canyon habitats reflected a more moist and cool environment. At river level near Stanton's Cave, where only desert and scrub species can survive today, a forest typical of higher elevations once flourished.[3]

In the century since Stanton and his group first attempted to establish a railroad route through the canyon, scientists have

Stanton's Cave, at River Mile 31, is almost 150 feet above the surface of the river. Named after Robert Brewster Stanton, who led a river expedition through Grand Canyon in 1890, Stanton's Cave is a well-preserved vault of information on the natural history of the Colorado River environs over the last forty thousand years. *Robert C. Euler*

The split-twig figurines found in Stanton's Cave are small fetishes, usually made from a single willow bough or other flexible streamside tree or shrub. These ceremonial animal figurines were made by the Archaic Pinto Basin peoples, the first known inhabitants of the Grand Canyon, about four thousand years ago. *Photo courtesy Museum of Northern Arizona (Photograph NA 9419Aa.1.2)*

accumulated a large database on the natural history of the
Colorado River. Modern science is documenting environmental
perturbations caused not only by changing climates of the past,
but by the human need to control the river in the present. When
John Wesley Powell dared the first two boating adventures
through the Grand Canyon in 1869 and 1872, certainly it was
beyond his wildest dreams that one day the Colorado River
would be regulated by two of the world's largest dams. But such
is the case. By 1936, Hoover Dam at the downstream end of the
canyon was transforming the lower reaches of the river into a
lake, backing still water into the lower Grand Canyon and
drowning many of its rapids. And a little less than three decades
later, Glen Canyon Dam at the upstream end was harnessing the
river's flows and altering its very nature.

Today the river in Grand Canyon is what Glen Canyon Dam
makes it. Its flow is regulated, usually through power-producing
turbines; its temperature is constantly cold; and its namesake,
the reddish-colored silt that once choked its flow, has been
replaced by flourishing green algae. The story of Glen Canyon
Dam and the Colorado River is a classic ecological tale. In this
telling the consequences of river control for the aquatic and
terrestrial ecology of the riverine environments in the Grand
Canyon are judged neither bad nor good. Some living systems
have been lost, but others have been gained.

The River System

The Colorado River is, without question, the premier
watercourse of the arid Southwest. From its glacier-fed origins at
elevations above 14,000 feet on the Continental Divide in Rocky
Mountain National Park, the Colorado River flows some 1,400
miles to Mexico and the Gulf of California. The river and its
tributaries drain parts of seven states and a small portion of
Mexico. The drainage basin, encompassing 244,000 square
miles, has been split for political and administrative purposes
into two smaller basins, with Lees Ferry the designated dividing
point. All of the drainage upstream of Lees Ferry, a total of
some 108,000 square miles, is known as the upper basin. The
portion downstream is called the lower basin.

The Grand Canyon essentially begins at Lees Ferry at the
approximate midpoint of the river's length, halfway between the
Continental Divide and the Gulf of California. The Colorado is
a large and powerful river by the time it enters the canyon — up
to 300 feet wide with a deep, strong current that averages more
than 8 million acre-feet (maf) per year. (An acre-foot is the

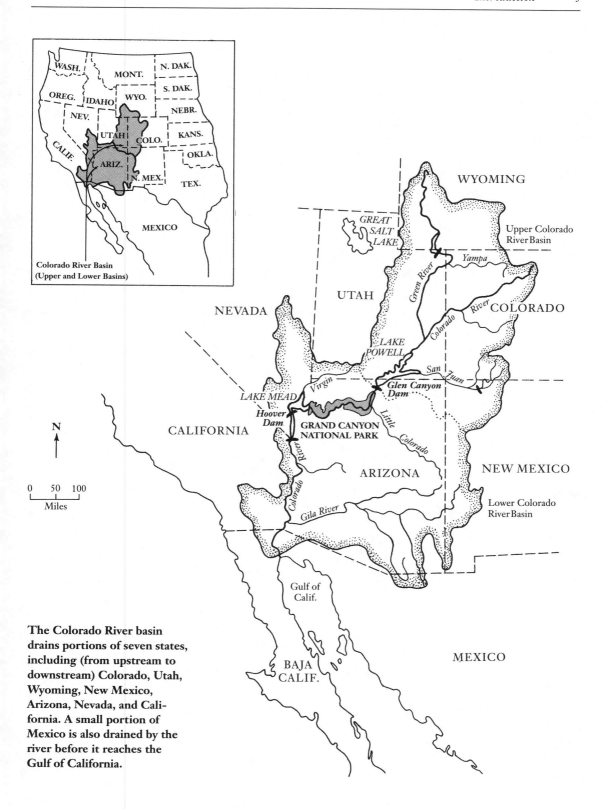

Colorado River Basin
(Upper and Lower Basins)

WYOMING

GREAT SALT LAKE

Upper Colorado River Basin

NEVADA

UTAH

Colorado River

COLORADO

Green River

Yampa

LAKE POWELL

San Juan

Virgin

Glen Canyon Dam

LAKE MEAD

Hoover Dam

GRAND CANYON NATIONAL PARK

Little Colorado

CALIFORNIA

ARIZONA

NEW MEXICO

N

Colorado River

Lower Colorado River Basin

0   50   100
Miles

Gila River

Gulf of Calif.

BAJA CALIF.

MEXICO

The Colorado River basin drains portions of seven states, including (from upstream to downstream) Colorado, Utah, Wyoming, New Mexico, Arizona, Nevada, and California. A small portion of Mexico is also drained by the river before it reaches the Gulf of California.

amount of water or sediment that would cover an acre of ground to a depth of one foot.)

By measure of volume alone, the Colorado is virtually complete by the time it reaches Grand Canyon. The majority of water supplied to the river comes from snowpack accumulated in the mountains of the upper basin. In contrast, the lower basin states contribute only a small portion to the overall flow. The river also has lost most of its elevation, approximately 75 percent, by that point. Enough drop remains, though, to permit formation of the rapids for which the Grand Canyon is so famous.

The entire 277-mile length of the Colorado River corridor through Grand Canyon, from Lees Ferry to the Grand Wash Cliffs at Lake Mead, is within the jurisdiction of Grand Canyon National Park. Add the 15 miles of river from Lees Ferry upstream to the dam (which falls within Glen Canyon National Recreation Area), and the total comes to 292 miles: the single longest protected stretch of canyon and river in the world. But the river is protected in more than just a legal sense, for it is the heart and soul of several million acres of wilderness isolated by the canyon's rugged topography. From Navajo Bridge at Marble Canyon to Hoover Dam, a distance of more than 300 miles, automobiles cannot cross from one side of the canyon to the other. For most of this length, the only way to gain access to the river is by foot, boat, or helicopter.

Hoover Dam was completed in 1935 to become the first river-control structure to modify the Colorado River in Grand Canyon. *U.S. Bureau of Recla-mation, Lower Colorado Region*

For hundreds of thousands of years the Colorado River flowed through the Grand Canyon undisturbed by man. Over the course of a single year, it followed dramatic but relatively predictable natural patterns. The result was a complex and finely tuned riverine ecosystem, a closely related and interacting collection of plants, animals, and environmental conditions. Fish and other aquatic life had adapted to the turbidity and seasonal fluctuations of a natural desert river. Plant and animal life on the riverbanks had become accustomed to the natural cycle of annual scouring floods. Although desert rivers are dynamic and subject to change, the living systems of the Colorado River in Grand Canyon had achieved a balance with the predictably harsh environment in which they had evolved.

This balance ended with man's control of the river. Hoover Dam profoundly transformed the lower few miles of the Colorado, but the most significant changes for the entire system came with the closing of the floodgates of Glen Canyon Dam in 1963. Once that dam became operational, a complete physical and biological alteration of the river in Grand Canyon was underway.

The Dams: Hoover and Glen Canyon

Hoover Dam, originally named Boulder Dam, was the largest dam in the world at the time of its construction. President Franklin D. Roosevelt dedicated the finished structure on September 20, 1935, before an audience of twelve thousand people. Effectively blocked for the first time in its history by the hand of man, the river immediately began to back up behind the dam, forming Lake Mead. By 1941, the reservoir stretched almost 110 miles upstream, extending well into Grand Canyon to the foot of Bridge Canyon Rapid at River Mile 235 (river miles are measured downstream from Lees Ferry, River Mile 0). The lowermost 43 miles of the Grand Canyon section of the river swelled into an artificial lake, one soon choked with silt and debris as the entering river slowed to a standstill.

The river upstream of Lake Mead continued to flow freely until March 1963, when Glen Canyon Dam was completed. Authorized by the Colorado River Storage Project Act of 1956, this dam was designed primarily for water storage, although some secondary functions were also identified. They included the enhancement of recreation and the downstream environment, production of hydroelectric power, and entrapment of the river's immense sediment load, which was filling Lake Mead at an alarming rate. Silt deposits were reducing the storage capacity of Lake Mead by more than 135,000 acre-feet per year, a rate that would have filled the reservoir within just a few hundred years.

The ultimate reason for the construction of Glen Canyon Dam was clearly political. The cornerstone of western water law — the concept of prior allocation (first in time, first in use) — dictated that there would be intense competition to see who would be the first to use the river's water. By the early 1920s, water in the developing Southwest had become too precious a commodity to overlook, and the seven states through which the Colorado River flowed wanted assurances that they would have access to their shares of the river's water. The Colorado River Compact of 1922 divided rights to the river among these states.

The Grand Wash Cliffs, in the foreground, are the western boundary of the Grand Canyon. Upper Lake Mead extends more than 40 miles upstream of the cliffs, inundating the river and transforming it into a reservoir. *Michael Collier*

Glen Canyon Dam is the largest single influence over the Colorado River corridor in Grand Canyon. Completed in 1963 to satisfy the water-storage requirements of the Colorado River Compact of 1922, the dam has dramatically altered the natural cycles of flooding in the river through Grand Canyon. *U.S. Bureau of Reclamation, Upper Colorado Region*

As a condition of the compact, the upper basin states were required to allow 75 maf to pass to the lower basin states every ten years (an average of 7.5 maf per year; a later treaty with Mexico raised the average annual release to the lower basin to 8.23 maf).

To control the distribution of water, the Colorado River Storage Act of 1956 called for the building of a large dam and reservoir very near the dividing line between the upper and lower basins. The dam was both to store water for the upper basin and to guarantee the accurate delivery of the mandated amount of water, on schedule, to the lower basin. The first contracts associated with the construction of Glen Canyon Dam were awarded almost as soon as the ink was dry on the act.

Roads were built to the site chosen for the dam, 75 remote miles from the nearest town of Kanab, Utah. Just over the Arizona state border, on a mesa formerly inhabited only by jackrabbits and sagebrush, the new town of Page rose up overnight to support the construction efforts.

The first concrete for the dam was poured on June 17, 1960. Three years later, on March 13, 1963, the 710-foot-high structure was ready to do its job, and the gates to the river-diversion tunnels were sealed off. Lake Powell, called by some the "Jewel of the Desert," began to fill. It was to have a total capacity of more than 27 maf, making it bigger than Lake Mead (at 25 maf) and one of the largest reservoirs in North America.

The "New" River in Grand Canyon

During the late 1960s, after Glen Canyon Dam had become a reality, a great lament arose from the budding environmental community. At that time the objection was over the inundation and destruction of Glen Canyon. Little, if any, concern was voiced that the operation of Glen Canyon Dam would inevitably affect Grand Canyon National Park, just a short distance downriver. Nearly unnoticed, the riverine ecosystem within the park began to reflect the impact of the dam almost as soon as it became operational.

The aquatic environment was the first to change, being instantly and thoroughly altered. Water entering the canyon no longer ran the muddy red of the ancient Colorado, but sparkled a clear green, completely free of sediments. It was deep-water cold, released from penstock gates 200 feet below the surface of Lake Powell. Summer and winter, the water temperature remained stable at about 48°F. The conjecture is often repeated that if Friar Francisco Garcés, the Spanish missionary who in 1776 viewed the river from the canyon rim and assigned it the name "Río Colorado" (river colored red), had seen the river as it appears today, he would probably have given it a different name — perhaps "Río Verde" or even "Río Esmeralda," the green or emerald river. To ecologists, the simple change in river color from red to green signals a fundamental difference in the very nature of the riverine ecosystem, a change no less profound than the sudden illumination of a darkened room.

Though unanticipated at the time, the clear river meant that the energy available from the sun would no longer reflect off the surface of the water, but penetrate deeply. The effect has been to increase the river's biological productivity significantly.

Introduced species of fish, especially the highly prized rainbow trout, and other nonindigenous forms of aquatic life soon colonized the clear and now-productive river. Native fish, already in decline from the pressures brought to bear by the introductions of carp, catfish, and other aquatic life, were poorly adapted to the perpetually cold, clear water and did not fare well.

The riparian, or streamside, habitat of the river corridor changed more slowly, but no less drastically. Control of the river eventually resulted in an increase in the amount of riparian vegetation growing along the river margins. As the dam stabilized the river's peak flows, annual floods that previously had scoured the riverbanks were eliminated, allowing a vigorous growth of new vegetation to develop right down to the river's edge. Because of the added vegetation, populations of native insects, lizards, toads, small mammals, and birds also flourished.

Paradoxically, even as the dam has made this new streamside abundance of life possible, its long-term existence is threatened by the new river. The vegetation is rooted in and dependent on pre-dam sediments deposited by the river. Now that its load of silt is trapped in Lake Powell, the river is primarily an erosive force, with little suspended sediment to replace the vast amount it carries away. One of the most far-reaching results of the dam has been a gradual decline and in some cases disappearance of sandbars, or beaches, available to support wildlife-producing, riparian vegetation. The possibility that the dam-controlled river could scour itself to bedrock, removing the newly gained wildlife habitat in the process, only became evident more than ten years after Glen Canyon Dam became operational.

Dam Control and Stabilization of the River-Running Industry

Early explorers and adventurers, when faced with the dangerous rapids and remoteness of the canyon, usually elected to avoid it altogether. Only one hundred hardy individuals had traversed the canyon by boat between Major Powell's first voyage in 1869 and the year 1949 — a time span of eighty years — and several had not lived to tell about it. But the development of more suitable river-running equipment, flow control by Glen Canyon Dam, and the evolving attitudes of American society changed all that. The availability of military-surplus pontoon rafts shortly after World War II transformed river rafting into a popular recreational activity enjoyed by thousands of people. The changing perceptions of an increasingly urbanized population came to place a higher value on the wilderness experience, particularly the "re-creational" benefits gained from voyaging down a wild river.

Use began to increase after 1950, so that by 1954 a total of two hundred people had run the river through Grand Canyon. More than three hundred fifty people ran it in 1962 alone. The numbers dwindled in 1963 and 1964 when the river was reduced to a trickle to fill Lake Powell, but the cold, clear, relatively steady flows that began in 1965 drew river runners in ever greater numbers. Use of the river skyrocketed. More than a thousand people traversed the canyon in 1966; more than sixteen thousand in 1972. Once remote and seldom visited, the river corridor had become a popular vacation destination for thousands of people. Yet accessibility has its price: what was at one time a true wilderness experience had become a recreational experience in a wilderness setting.[4]

The exponential increase in recreational use of the river corridor alarmed many of the river runners and National Park Service officials. Fearing the effects of overcrowding, the park held use at 1972 levels and instituted a major research program to study the interrelationships between human use of the river and its natural resources. The Colorado River Research Program, lasting from 1973 to 1976, was directed by R. Roy Johnson of the National Park Service. The program addressed the problem through the combined efforts of twenty-nine separate scientific studies.[5] Subjects ranged from riverine ecology and water quality to the sociological impact of increased recreational use on the river runners themselves.

The studies found congestion and associated environmental disruption, much of it due to the large number of people who wanted to travel down the river at the same time: during the warm summer months that coincided with their annual vacations. To compound the problem, these visitors were camping primarily at a few, regularly used beaches. Recreation-related damage such as soil compaction, wildfire, and vegetation disturbance were concentrated in these areas, as were the associated litter, camp garbage, campfire charcoal, and human waste. This damage, in turn, was disturbing the natural distribution and abundance of plants and animals. Furthermore, regulated flows from the dam were no longer cleansing and replenishing the beaches as spring floods had before the dam was built.[6]

To mitigate the congestion and environmental impact in the new, controlled river corridor, the National Park Service imposed strict regulations on river running. Scheduling, use

Concern over the impact of heavy recreational use of many camping beaches in the early 1970s led to scheduling and regulation of white-water boating by the National Park Service. By the early 1980s, beach erosion and other changes caused by fluctuating flows from Glen Canyon Dam had greatly exceeded the problems caused by recreational impact. *Dugald Bremner*

limits, and the reduction of human-influenced disturbances became the new park management priorities. River parties were assigned specific launch dates to minimize overcrowding; river runners were required to carry out all trash, garbage, and solid human wastes; and restrictions were placed on fires and the burning of driftwood, now a nonrenewable resource. The new regulations to protect the environment were quickly accepted by most river runners, who shared the concern for the canyon's environment and, more than anyone, recognized the threat of overuse to the quality of a very special place and the experience it provided.

The Flood of 1983

An unforeseen time bomb was ticking away during those first two decades of the new river era, when Lake Powell was filling and park management attention was focused on the problems of recreational influence. Little thought had been given to the potential for downstream floods once the lake was full. Unlike the managers of Hoover Dam, who were required by law to prevent downstream flooding and degradation of resources and property, Glen Canyon Dam managers had no such mandate.

A temporary water release through the dam's river outlet tubes for a few days in late June 1980 was the only signal that marked the filling of Lake Powell. After that, the Bureau of Reclamation attempted to maintain the reservoir at as high a level as possible as a hedge against drought and to maximize revenues from hydroelectric power generation. The upper basin states also desired to retain as much of their share of river water as the 1922 compact allowed, even though they were unable to make full use of their allotment. By 1980, the upper Colorado River basin was regulated by many dams and river-control structures for this and other purposes. The Bureau of Reclamation gambled that a system of upstream reservoir management and runoff-prediction technology was enough to prevent too much water from entering Lake Powell.

The bureau's strategy and luck were to last for less than three years. A higher-than-average snowpack in the Rocky Mountains in the spring of 1983, an unusually rapid snowmelt, and unseasonal rainfall in May and June of that year combined with full reservoirs throughout the upper basin were to have profound consequences for the Colorado River in Grand Canyon. The resulting flood of June 1983 was unprecedented for the river corridor environment that had developed under controlled, stable conditions since 1963. Through sediment loss and vegetation scouring, more long-term damage was done to the developing river ecosystem in one brief month of virtually uncontrolled flooding than had been done by all the recreational activities of the previous twenty years combined. Clearly, the 1983 flood proved that operation of the dam was the largest single influence over the fate of the river corridor.

The influence of the dam can swing both ways. The hand of man, and not nature, controls the operations at Glen Canyon Dam. The same hand that permitted the disastrous 1983 flood might be swayed, albeit with difficulty, to manage the dam for the benefit of downstream resources as well as for water storage and power production.

The Colorado River in Grand Canyon has become a test-tube experiment on an immense scale. Much like a carefully tended aquarium, the river is now effectively sealed off by dams and reservoirs at both ends, and the input of water and nutrients is strictly controlled and monitored. The outcome of the experiment will be determined by the Bureau of Reclamation and the manner in which it releases water from Glen Canyon Dam. The extent of the water released at any given time, its temperature, patterns of daily and seasonal discharge fluctuation, and the management of flood flows all have a substantial effect on the river's living and nonliving systems. A single-minded pursuit of hydroelectric power generation and ever-more water storage will eventually degrade the river environment in Grand Canyon. Slight changes in how the dam is operated, however, could actually enhance downstream natural resources.

Glen Canyon Environmental Studies

The Glen Canyon Environmental Studies has been the Bureau of Reclamation's most ambitious response to the problems facing the river's future. Authorized by Secretary of the Interior James Watt, the studies were organized as a cooperative venture among several government agencies: the National Park Service, Bureau of Reclamation, U.S. Fish and Wildlife Service, U.S. Geological Survey, and the Arizona Game and Fish Department. The goal was to analyze the effects of the dam on downstream resources. The investigations quickly became the single largest and most costly ecological research project ever undertaken by the United States government up to that time; more than seven million dollars was spent searching for the secrets of the river corridor from 1983 to 1987.

The Glen Canyon Environmental Studies introduced the concept of managing the dam for the benefit of plants, animals, and river running in Grand Canyon. This concept and its alternative management options almost immediately ran into a stone wall of legal and political opposition from agencies and organizations that benefited from the dam's current operations.[7] The beneficiaries of the status quo saw little advantage in nurturing either natural or dam-altered riverine ecosystems.

The Colorado River of the West has become a single, vast, complex plumbing system, with Glen Canyon and Hoover dams the controlling faucets. In literature, rivers often symbolize the passage of life; in the arid Southwest, the Colorado River *is* life for millions of people and empires of agriculture, manufacturing, mining, development, and recreation. It is little wonder, then, that any attempt to alter the status quo at Glen Canyon Dam would be met with resistance. As Philip Fradkin noted in *A River No More*, "the Colorado is the most used, the most dramatic, and the most highly litigated and politicized river in this country, if not the world."[8]

Lawsuits pertaining to Colorado River environmental concerns and water allocation have become commonplace, with much of this legal action focusing on the operation of Glen Canyon Dam. And there is no end in sight. An ever-increasing population in the Southwest presages escalating requirements for Colorado River water and more lawsuits in the future. Conflict between calls for environmental protection of the river in Grand Canyon and demands for water and power revenues has led to a complicated set of criteria for managing this controversial resource. The pendulum of biopolitical management appears to be swinging to the side of resource protection. Nevertheless, the final outcome will be hotly contested in the courts for decades to come.

Together, humans and nature have brought about the perturbations that are the trademark of the post-dam river. The river corridor has been constantly, but slowly, evolving throughout the millions of years it has existed. Humans have recently accelerated this natural process of change and, at least temporarily, altered its direction. The river of the future will undergo still further modifications that are only beginning to be predicted. Natural history, the science of describing the living and nonliving components of nature, is one vehicle from which we can gain perspective on this change.

In the following chapters we review what little is known of the river environment of the pre-dam era and explore the dynamic and precarious condition of the present ecosystem. And most importantly, we glance at the future condition of this unique place. There is no question that the aquatic and terrestrial habitats found along the river corridor in Grand Canyon will decline or flourish as a function of the manner in which water is released from Glen Canyon Dam. Decisions made on environmental priorities within the national park and on economic priorities relating to water management throughout the vast drainage of the West's most important river will determine the future of the Colorado River in Grand Canyon.

PART I

Water Meets Rock:
The Physical Environment

THE RIVER
Dramatic Rise and Fall

On the evening of September 18, 1923, a small group of river runners pulled their boats into the left bank of the Colorado River and prepared to make camp for the night. They belonged to the Birdseye expedition, a U.S. Geological Survey group charged with mapping the Colorado and locating potential damsites along its course. In the early 1920s many interests in the West were eager to harness the wild waters of this most unpredictable of rivers. Flood control for downstream communities, water storage for irrigation and development, and hydroelectric power generation all depended on the construction of dams. The Grand Canyon, with its narrow, deep gorges, offered obvious possibilities. By the evening of the 18th, the surveyors of the Birdseye group had already identified dozens of feasible damsites in the 179 miles of the Grand Canyon they had traversed so far.

Guided by a veteran Colorado river runner, Emery Kolb, the party had encountered few difficulties thus far in the trip. But that was soon to change. Because they had lost use of their radio, the expedition's members missed broadcasts advising that an extraordinarily heavy rainstorm had drenched the basin of the Little Colorado River, a tributary 120 miles upstream. Normally a docile stream, the Little Colorado had swollen to a torrent and was sending its brown floodwaters down the Colorado toward the unsuspecting boatmen. Warnings that the expedition should look for high ground went unheard. In his account of the subsequent events, Lewis Freeman, a member of the party, wrote: "If such warnings had been received, we would have selected a broad open section, with ample room to back away from a rise, and waited for the flood to pass. Unwarned, we were surprised at a time and place far from favorable — twilight on the brink of Lava Falls."[1]

To this day, a Colorado River guide's nightmares as often as not invoke the specter of Lava Falls, a maelstrom of foam, black rocks, and standing waves. Kolb was to experience a living nightmare that fateful September night. Unaware of the flood that was surging downriver toward them, he chose to camp on a narrow stretch of beach at the head of the rapid. He had the group unload its bedding and kitchen gear but decided the mooring was bad for the boats. Just before dark, Kolb directed the crew to line the boats past the worst of the rapid to what Freeman called "a narrow crescent of a beach below."

Nightmare at Lava Falls

No sooner had they completed this back-breaking chore and begun the trek back to camp in the enveloping dusk, when they noticed the river level coming up — fast. Realizing that the moored boats could be swamped and swept away, the men scrambled back over the rocks to reposition them. Again and again they found new moorings, but the water kept rising, driving them before it. Their one empty boat, the one that had held the camping gear, was hoisted up the ledges with block and tackle. But the remaining boats were still laden with their precious equipment and could not be lifted. Eventually, there was nothing to do but launch themselves and their boats into the black night and swift waters, trusting their skills and luck to find safe haven downstream. Saving the boats took most of the crew's strength but was not the only problem. Bedding and kitchen gear on shore had to be dragged up the rocky talus slope repeatedly to ever-higher ground above the swelling current. Trapped between cliff and river, the crew spent the entire night moving boats and stumbling back and forth along the rocky slope to save equipment strung the length of the rapid, all to the deafening roar of Lava Falls. Morning found the men exhausted and soaked. The river had risen 13 vertical feet that night and was to rise 8 more feet before beginning to subside.

The flood eventually reached a peak estimated to be well over 100,000 cubic feet per second (cfs).[2] The Bright Angel gauge near Phantom Ranch measured 98,500 cfs for the date, but with the heavy storms it very likely underestimated river flow 92 additional miles downstream at Lava Falls. The expedition members rested below Lava for three days waiting for the flood to recede. Even though they did not have the benefit of their radio, they assumed correctly that the flood had originated in the malodorous, salty Little Colorado River: "A distinctly unpleasant odor from the water suggested that the flood was coming from the Little Colorado River. Mingled with this was a humid, almost tropical, smell — rank vegetation and a suggestion of the perfume of flowers."[3]

That dramatic summer flood of 1923 illustrates a characteristic of the free-flowing Colorado: it was a river of

Figure 1.1. Members of the Birdseye Expedition assembled at Lees Ferry in July 1923. *From left to right:* Leigh Lint, boatman; H. E. Blake, boatman; Frank Word, cook; C. H. Birdseye, expedition leader; R. C. Moore, geologist; R. W. Buchard, topographer; E. C. La Rue, hydraulic engineer; Lewis R. Freeman, boatman; and Emery C. Kolb, head boatman. *L. R. Freeman (Topography D#9), Courtesy U.S. Geological Survey Photo Library, Denver*

Figure 1.2. A flash flood on the Little Colorado River caused the Colorado River to rise to more than 100,000 cfs in late September 1923. Members of the Birdseye Expedition, camped at Lava Falls, labored all night to keep their boats and equipment above the rapidly rising water. Here one of the expedition's wooden boats is tied to the shore at the head of Lava Falls the following day. *E. C. La Rue (#604), Courtesy U.S. Geological Survey Photo Library, Denver*

extremes. The radical change in flow that trapped the Birdseye party at Lava Falls was the result of relatively local conditions, an unusually intense storm over the Little Colorado River drainage. Because no dams of consequence have been built on the Little Colorado River since 1923, a comparable flood could happen again; local storms still wreak havoc on the river. But their overall effect is minor compared to the seasonal floods that once swept down the mainstem of the Colorado each spring.

Periodicity of Flow

Before Glen Canyon Dam was built, river flow through the Grand Canyon fluctuated dramatically but predictably over an annual cycle: low in winter and high during the spring and early summer. Extremes of water flow ranged from a record low of about 700 cfs in December 1924 to an estimated peak of more than 300,000 cfs during the spring of 1884. Normally, the average annual high flows were in excess of twenty times the average low flows. Records from the river gauge located near Phantom Ranch indicate that the average annual maximum flow was between 85,000 and 95,000 cfs. The average minimum was about 4,000 cfs. Every ten years or so, a flood of at least 120,000 cfs ripped through the canyon, with the largest flood for the period of gauging station records reaching 200,000 cfs in the early summer of 1921.

Figure 1.3. The completion of Glen Canyon Dam in 1963 marked an end to the huge annual floods that flowed through Grand Canyon. Annual peak flows of the river were dramatically reduced as a result.

The phenomenal flood of 1884 was more than sixty times the average low flow. Some observers speculate that this unusually high flood was related to a temporary worldwide climatic change brought about by the eruption and eventual explosion in 1883 of Krakatoa, a large volcano in the East Indies. The ash produced from the eruption spread across the globe, partially blocked the sun, and briefly created a mild version of the conditions we have come to refer to as a "nuclear winter." This atmospheric change may have caused an unusually deep snowpack in the mountains of the Colorado River basin and the subsequent high runoff. The size of the flood cannot be known for certain because it predated the installation of the gauging stations, but a reliable estimate has been made on the basis of high-water debris left above the known 200,000-cfs flood in 1921.

Glen Canyon Dam has virtually eliminated these extreme seasonal highs and lows, imposing instead daily fluctuations. Under most circumstances, water is released from the dam in response to the varying demand for hydroelectric power in cities far from Grand Canyon or to fulfill downstream water-delivery requirements. Flows are now somewhat predictable, ranging

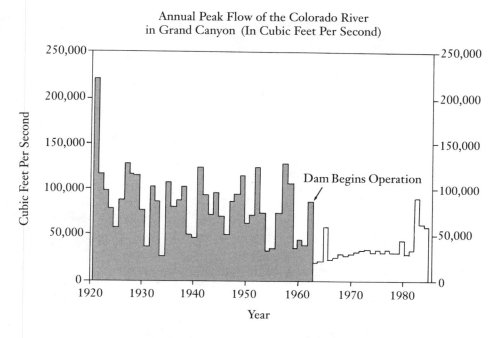

Annual Peak Flow of the Colorado River
in Grand Canyon (In Cubic Feet Per Second)

from as little as 1,000 to 3,000 cfs for the daily low to as much as
31,500 cfs for the daily high.

Since the need for electricity dictates daily water releases,
flows also exhibit something of a seasonal pattern. Spring and
fall peak releases are usually less than those in midsummer and
winter (energy needs for cooling and heating), although the
variation does not approach pre-dam extremes. The average
annual high flow has been significantly reduced to about 30,000
cfs. The average ten-year flood, from 1963 (when the floodgates
of the dam were closed) to 1983, was only about 40,000 cfs. In
1983 a flood discharge of 92,200 cfs approximated pre-dam
annual high flow and signaled the end of one era and the
beginning of another in Glen Canyon Dam's influence on
downstream environments.

Dam Operation

The reasoning behind the operational scenarios of the dam are
complex and directly related to several factors, including legal
requirements, annual precipitation in the upper basin,
hydroelectric energy contracts, reservoir levels, and to a much
lesser extent, recreational river-boating requirements. Glen
Canyon Dam is the key regulatory structure for the
management of the upper basin water distribution into the lower
basin states and Mexico. Today, flows through the Grand

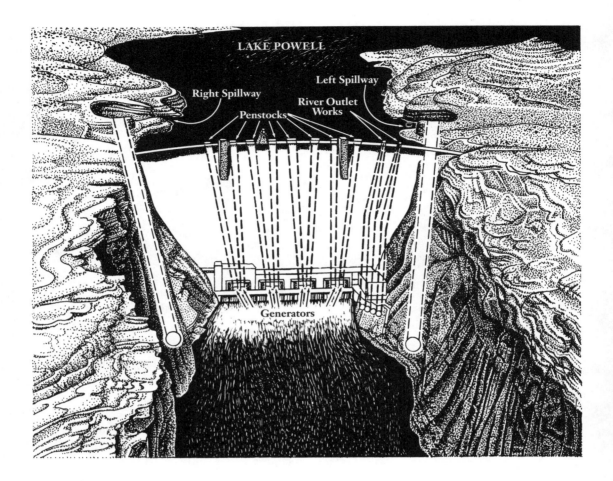

Canyon are governed by storage and release decisions that
determine annual, monthly, hourly, and minute-by-minute water
releases.

A discussion of historical releases from the dam can be divided
into four distinct periods of operation: (1) the filling of Lake
Powell, 1963–80; (2) the full-lake operation, 1980–83; (3) the
lake overflow and subsequent flood, 1983; and (4) flows in the
post-flood era, since 1983.

The Filling of Lake Powell, 1963–80

Construction of Glen Canyon Dam, which began in 1956, was
completed in March 1963 with the closing of the floodgates. For
the next seventeen years, water releases from the dam were
governed by policies known as the *filling criteria* and the
operating criteria. The management strategy was to fill the

9-trillion-gallon (27 maf) Lake Powell as rapidly as possible, a goal initially constrained only by the need to minimize impact on the downstream operation of Hoover Dam.

For the first two years of dam operation, from 1963 to 1965, only 5 maf of water was allowed to flow through the canyon, a reduction of about 75 percent of normal flow. The intent of the minimum releases was to raise the reservoir's water level to the minimum elevation, called the *power head*, necessary to drive the generators and produce electricity. The consequence, however, of achieving a power head within the first two years at Glen Canyon was to deplete Lake Mead's storage downstream to a level below Hoover Dam's capacity to produce power. To restore this capacity, nearly 11 maf of water was released through the Grand Canyon in 1965. Throughout the filling period, releases rarely exceeded 31,500 cfs, the maximum amount of water that could be passed through Glen Canyon Dam's eight generators.

It was during the nearly two decades of reservoir filling that the downstream riverine environments began to adjust to dam-controlled flows. Annual floods ceased, and water flow began to assume a predictable pattern. At the dam, water releases roughly followed a peaking schedule of 8 A.M. to 5 P.M. As municipal and industrial energy customers began their workday, power-producing water was released from the lake; when energy demands decreased at the end of the workday, discharges from the dam decreased accordingly. Weekend water releases were typically very low, also following the pattern of human energy consumption. Over the weekend, water releases fell to a minimum, and it was often Monday morning before significant releases were resumed.

After Lake Powell filled in 1980, water releases through the power plant generally continued to follow this regimen. River guides learning to navigate the river find they have to keep track of the daily "tides." The farther one floats downstream of the dam, the more difficult it is to predict the almost constantly changing water levels. River parties rarely launch from Lees Ferry in the early morning because the water is still low from the nighttime discharges. Traveling at about 5 miles per hour, an 8 A.M. release usually does not reach Lees Ferry until 10 or 11 A.M. The first rapid of any consequence is Badger Rapid, 8 miles downstream of the launch point. It is always rocky in low water. River parties launching on a weekend, especially during low-water years or during fall and winter, face a very rocky river until the Monday morning releases raise the flow.

Figure 1.4. Water from Lake Powell may pass through Glen Canyon Dam in three ways, depending on the level of the lake and inflow from the Colorado and San Juan rivers upstream. Water is normally taken in through the pen-stocks and passes through the generators in the power plant to produce electricity before being released into the river below. When both the lake level and inflow from upstream are high, surplus water may be released through the river outlet works and the spillways.

As a general rule of thumb, it takes water releases from the dam about twenty-four hours to reach Phantom Ranch, a little more than 100 miles downstream. In other words, a Monday high-water release does not reach the Phantom Ranch area until Tuesday, and it will not reach Lava Falls until late Wednesday. Every capable guide monitors the water level, attempting to predict how much water will be in the river when it is time to navigate the largest rapids.

Full-Lake Operation, 1980–83

Once Lake Powell had filled in 1980, reservoir management strategy became more complex. Decisions governing releases from the dam were directed by the operating criteria. These criteria were supposed to take into consideration all the uses and diverse users of the reservoir systems on the Colorado River, including flood control, water-quality control, recreation, and other issues. Nevertheless, the overriding objective was, and continues to be, to keep the reservoir as full as possible as a hedge against drought, yet leave enough room in the lake to accommodate the spring and summer runoff, all while maximizing power production.

The big guessing game takes place each spring when the Bureau of Reclamation must anticipate how much runoff will come into the lake from the upper Colorado and San Juan rivers. In other words, dam operators must determine in advance what Mother Nature is going to do so they can draw down the reservoir level just enough to accommodate the expected inflow. They receive their climatic data from the National Weather Service, frequent snowpack data from the Soil Conservation Service, and streamflow data from the U.S. Geological Survey.

The Lake Overflow and Flood, 1983

During the summers of 1980, 1981, and 1982, everything worked according to plan. The weather cooperated, and average amounts of precipitation fell throughout the upper Colorado River basin. But the winter of 1982 to 1983 was one of unusually heavy snowfall, and by January 1, 1983, with Lake Powell at the desired level for that date of 22.6 maf, forecasters were predicting an inflow 112 percent of normal. Confident that they were still in control of the situation, dam operators released higher than normal amounts of water from Glen Canyon and Hoover dams, even allotting water-hungry Mexico 200,000 acre-feet more than the previous year. But the increased releases were not enough. Because of unexpected late spring snowstorms, forecasts for the next several months shifted ever upward, and inflow continued to increase beyond predictions.

By June 1, the forecasts were predicting an inflow of 131 percent of normal, and discharges from Glen Canyon Dam were increased to 28,000 cfs over twenty-four hours a day. Releases were still routed through the turbines, but dam managers were beginning to admit defeat in the Lake Powell inflow guessing game. Water was pouring into the lake faster than it could be released. The situation escalated from serious to critical and reached emergency proportions by the beginning of summer. June 1983 proved to be a month few of those involved in Grand Canyon river management will ever forget.

In an effort to control the level of the lake, Bureau of Reclamation officials began releasing the maximum amount of water possible, 31,500 cfs, through the power plant around the clock. Then, despite their reluctance to bypass the turbines, they were forced to open the dam's four river outlet tubes to their full capacity of approximately 17,000 cfs. Even that was not adequate, however, and officials took the action of last resort. They opened the giant, concrete-lined spillways that tunnel through solid rock on either side of the dam. With a combined capacity of 200,000 cfs, the spillways should easily have accommodated the overflow. They did not.

With continuous use, even at minimum discharges, the spillway flows began to surge erratically. Then water from the left (as one faces downstream) spillway darkened with chunks of red Navajo Sandstone, obviously gouged from the spillway walls. Distressed at the implication of these developments, the engineers shut down one of the spillways long enough for a quick inspection, and their worst fears were realized — the spillway walls were disintegrating as a result of cavitation. This process occurs when water flowing rapidly over a slightly roughened surface develops tiny vacuum pockets. In the spillways, wherever the vacuum pockets appeared, flakes began to exfoliate from the concrete lining. Microscopic particles came off first; then as the surface roughened, the process accelerated, and the particles progressively grew larger until they reached the size of boulders. The spillways were coming apart.

Continued use of the tunnels would only exacerbate the problem, but the Bureau of Reclamation had no alternative. Inflow was still increasing, and Lake Powell was rising ever closer to the top of the dam. The damaged spillway was reopened, but sustained releases were limited to 20,000 cfs on each side, with maximum peak releases held at 32,000 cfs.

Tom Gambel was the Bureau of Reclamation engineer in charge of dam operations during the crisis. Imagine the feeling in the pit of his stomach when he was faced with a full lake, inflow several thousand cubic feet per second above outflow, and disintegrating spillways. On June 28, releases reached 92,200 cfs (the highest since the dam was constructed), while river inflow to Lake Powell was cresting at 116,000 cfs; the dam was in danger of overtopping. Something had to be done.

One of the engineers came to the rescue with a successful, if implausible, solution: plywood. The engineers increased the depth of the lake by four feet, almost 1.5 maf of lake capacity, by installing 4-by-8-foot sheets of the best plywood money could buy on top of the spillway gates. Incredibly, this simple idea saved the day, and possibly the dam, although the bureau maintains that at no time during the flood events of June 1983 was Glen Canyon Dam in danger of failing. In early July the plywood was replaced with steel flashboards.

Lake Powell was 4.5 feet above the spillway gates in early July, but inflow was decreasing and the crisis abated. When it became possible to reduce the level of the lake through power-plant releases alone, both spillways were shut down and inspected. The damage was severe. The 3-foot-thick concrete walls were completely missing from large sections of the tunnels, steel reinforcing bars were impossibly mangled, and a series of great pits had been eroded out of the underlying sandstone. The largest hole measured 30 feet deep by 150 feet long.

In a frantic race against time, for the spillways might be needed again the next spring, the bureau let contracts to repair the damage. The design that allowed cavitation to take place was corrected by constructing an air slot 4 feet deep and 4 feet wide into the spillway tunnels. Apparently, entrainment of air into the spillway flow allows the water rather than the concrete to absorb and dissipate damage-causing shock waves. The spillway repairs and installation of the air slots were completed by late summer of 1984. On August 12, 1984, the spillways were tested and, at a flow of 50,000 cfs, cavitation failed to occur. Should the Bureau of Reclamation ever again need to release massive quantities of water, the spillways should hold.

The Influence of Gradient

Dam operators may not have achieved full control over the Colorado River, but it is they, not Mother Nature, who usually determine how much water flows downstream through the Grand Canyon and when. This is not to say that the Colorado is

Figure 1.5. Workers repaired
the spillways of Glen Canyon
Dam after the massive cavitation resulting from their
use during the flooding
emergency of late June 1983.
The worst damage was to the
left spillway, seen here, where
cavities measuring 30 feet
deep and 150 feet long required approximately 2,500
cubic yards of concrete to
repair. *U.S. Bureau of
Reclamation, Upper Colorado
Region*

a tamed river. Far from it. Anyone who has run the Colorado
knows it is a turbulent stream, marked by swift current, plunging
rapids, troughs and waves, swirling eddies, whirlpools, and
mysterious deep-water currents that boil to the surface in
seemingly placid stretches. Much of this turbulence can be
attributed to the river's gradient (the steepness of its fall), the
nature of the rock through which it passes, and debris flows that
have moved into the river channel by the many steep side
canyons along its course.

The Colorado drops approximately 1,900 feet in about 260
river miles between Lees Ferry (elevation 3,100 feet) and Lake
Mead (1,200 feet). This difference in elevation is coincidentally
illustrated by the height above the river of Diamond Peak, a
prominent landmark signaling a popular take-out point for river
runners. The top of this peak reaches an elevation similar to that
of Lees Ferry, 224 river miles upstream. The average loss in

elevation of the river, or the river gradient, is only 8 feet per
mile. This relatively modest figure suggests a gentle river. Such
would be the case if the entire drop were evenly distributed
along the course of the river, but 50 percent of the elevation
drop occurs in rapids that make up only 10 percent of the total
distance.

In hydraulic terms a river like the Colorado in Grand
Canyon, which loses most of its elevation in large rapids
separated by long flat-water pools, is described as a *pool-drop
river.* The overall channel gradient is dominated by the fall in
rapids, so that any change in the nature of the rapids has an
effect on the gradient. Riverbeds exhibit a natural tendency to
reach a constant gradient over time. This tendency is
counterbalanced by the unequal resistance of different rock types
to erosion, as well as by obstructing debris brought in by
tributaries. Resistant rocks and tributary inflow debris are
intricately linked in the formation of the Colorado River's most
famous element, the rapids.

Rapid Formation

The type of bedrock present at river level controls river-channel
characteristics such as channel width, the frequency of sand and
gravel bars, the fall of the river, and to a lesser extent, the
frequency of rapids. Shale, siltstone, and some sandstones are
relatively soft, quick to weather and erode. Where the river goes
through formations composed of such rocks (e.g., Bright Angel
Shale, Hermit Shale, and Dox Formation), the channel is
broader, less steep, and has relatively fewer rapids. Other
formations, composed of harder sandstones, limestone,
dolomite, and basalt (e.g., Coconino Sandstone, Redwall
Limestone, and Cardenas Lavas), are intermediate in their
resistance to erosion.

The hardest rocks in the canyon are also the oldest: the
ancient schists and gneisses laced with granite found in the inner
gorges — the deepest recesses of the canyon. They not only are
responsible for a narrow channel with a higher gradient, but
display an abundance of fault zones that have allowed the
formation of numerous tributary canyons. The Grand Canyon's
greatest density of large rapids occurs in the Upper Granite
Gorge. Here the frequency of major rapids is almost two-and-
one-half times as great as along other sections of the river. John
Wesley Powell was well aware of how the presence of gneiss
forewarned of an increasing level of difficulty of upcoming
rapids:

Figure 1.6. The summit of
Diamond Peak at River Mile
225 rises more than 1,900 feet
above the Colorado River,
approximately the same
difference in elevation exhibited by the river between
Lees Ferry and Lake Mead.
Bryan Brown

Figure 1.7. The Colorado
River has cut through
approximately 5,000 feet of
geologic strata to form the
Grand Canyon.

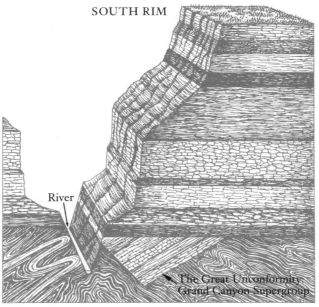

SOUTH RIM

Kaibab Limestone
Toroweap Formation
Coconino Sandstone
Hermit Shale

Supai Group

Redwall
Limestone
Temple Butte Limestone
Mauv Limestone

Bright Angel Shale
Tapeats Sandstone

Vishnu Schist and
Zoroaster Granite

River

The Great Unconformity:
Grand Canyon Supergroup

August 14. — At daybreak we walk down the bank of the river, on a little sandy beach, to take a view of a new feature in the canyon. Heretofore hard rocks have given us a bad river; soft rocks, smooth water; and a series of rocks harder than any we have experienced sets in. The river enters the gneiss! We can see but a little way into the granite gorge, but it looks threatening.[4]

Why rapids occur where they do has been a topic of much speculation. Some of the earliest work on the formation of rapids in the Grand Canyon was done in the 1960s by Luna Leopold, son of conservationist Aldo Leopold. Luna is well known for a keen appreciation of the natural processes that shape river drainages and the general topography of the planet. After a float trip through the Grand Canyon, he was convinced that the river was working toward a uniform gradient, but that it was in what he called a state of "quasi-equilibrium" that dictated a more-or-less regular alteration of flat-water pools and steep rapids. Leopold believed that the alternating pools and rapids were the consequence of a natural process that automatically maintained itself to achieve the minimum work expended in getting water from one point to another.[5]

Leopold saw riverbed erosion as one might see a carpenter sawing a board. The work proceeds easily until the carpenter hits a knot or a nail, at which time all the work of the saw is concentrated on the obstruction and little on the wood, until the obstruction is cut through. This illustration is analogous to the erosive force of a river in its channel. The channel tends to be eroded toward a uniform downstream gradient until reaching an irregularity such as a pile of boulders or layer of resistant rock. The work of the river concentrates on the obstacle until it is removed, and then the tendency toward a uniform gradient resumes.

Leopold's description of how flowing water seeks an equilibrium in a constant gradient makes sense theoretically, but it was the work of two University of Virginia scientists in the late 1970s which shed practical light on the true origins of Grand Canyon rapids. Alan Howard and Robert Dolan demonstrated that most rapids are created where the river crosses local and regional fault lines and fracture zones.[6] Drainages tend to follow these faults and fractures, eventually carving tributary canyons. Streamflow down the canyons carries boulders and other sediments to the river, forming alluvial fans. These in turn constrict the river into a narrower channel and form rapids. Given this observation, one might expect a rapid at the mouth of every tributary. Such is not the case. The junctions of the three largest tributaries in the Grand Canyon (the Paria River, the Little Colorado River, and Kanab Creek) have not produced substantial rapids, just small rapids or riffles. That is because the relatively low gradients of these tributaries do not have the drop necessary for the transport of large boulders to the river. Even a high-gradient tributary, however, does not always create a large rapid at its confluence if the bedrock at river level is soft and erodible. In these cases the immediate river gorge is usually much wider, the river is broader, and even the largest of the alluvial boulder fans may not constrict the channel sufficiently to form major rapids. Furthermore, soft bedrock tends to produce smaller debris to be transported into the river, debris that the river can more easily transport away.

Another condition that limits the occurrence of large rapids is the presence of resistant, but unfractured and unfaulted, bedrock at river level. The best example of this situation is the Muav Gorge from River Miles 140 to 160, where Upset is the only rapid of consequence. The resistant Muav Limestone that occurs at river level in this stretch is relatively unfractured; steep drainages entering the river are rare, and debris fans that could constrict the river and create rapids are not substantial. The frequency of rapids in this 20-mile section is only one-tenth of that for the overall river.

An important feature of the tributary delivery of debris to the mainstem river is that most tributaries have a much steeper gradient than the river. Because of this gradient difference, tributaries can deliver huge boulders to the river, boulders that cannot be removed by the river current except during the most extreme floods. Since Glen Canyon Dam became operational, eliminating or reducing annual, ten-year, and one-hundred-year floods, opportunities for the river to remove larger debris rarely exist. These consequences have foreboding implications for river running in the future. Many rapids have already become noticeably more severe. The message for river runners is that the river will become increasingly more difficult to navigate, and someday a side-canyon flood may stop safe passage altogether.

Massive rock falls or slides not associated with tributary canyons can also constrict the river and cause rapids.[7] Outstanding examples of this process can be seen at 10-Mile Rock, the Boulder Narrows (River Mile 18), and Randy's Rock (River Mile 126). Similar rock falls have occurred throughout the canyon; more can be expected. When they will come, how large they will be, and what effect they will have on river navigation is the subject of much animated conjecture.

Rapid Behavior

As we have seen, rapids in the Grand Canyon are usually created when debris delivered by steep tributaries blocks the river. Water backs up behind the obstruction to form a pool, spills over the boulders with great velocity and turbulence in the drop, then resumes its more gradual descent at the tail of the rapid. The behavior of the river just above, in, and below rapids is a repetitive pattern directed by the laws of physics. It is usually described as the *pool-rapid-runout sequence.*[8] Because the debris acts as a dam over which the water must fall, the water's surface elevation is higher at the head of the rapid than at its tail, or lowest point. The severity of the rapid is normally a function of this difference in elevation. For example, Sheer Wall Rapid at River Mile 14, a gentle little riffle, has a total drop of 8 feet. Lava Falls, on the other hand, the most feared of them all, drops 15 feet over a distance of 200 yards.

Typical Colorado River rapids share certain features. First, the approach to a rapid is characterized by the ponding, or slowing down, of the river as it backs up behind the constriction. Experienced river runners refer to the slow-moving water above Lava Falls and Crystal Rapid as Lava Lake and Lake Crystal, respectively. Robert Brewster Stanton, in his 1889 river expedition, quickly caught on to the relationship between the deceptively quiet pools and the rapids that followed: "At 4:00 P.M. we embark and sail about ¾ of a mile through one chute of very swift water, and then into a bay of glassy smoothness which of course means another rapid. At 4:10 we reach it. Large and terrific with huge boulders all over the upper part."[9]

Leaving the pool at the head of the rapid, the water accelerates into the constriction as a smooth but swift V-shaped trough, called the *tongue* of the rapid. At the bottom of the rapid, the water, still accelerating, streams from the constriction like a blast from a giant hose, forming the tailwater, or runout. Between the pool and the runout, high current velocity and obstructions in the rapid generate extreme turbulence. Reduced to the most basic principles, the behavior of the rapid is a function of the amount of potential energy stored in the deep pool above the rapid constriction which is converted to hair-raising kinetic energy as the water surges down through the constriction.

Enormous amounts of energy are gathered and released in a variety of wave forms. These waves develop in characteristic places. Nonbreaking or lateral waves bind the tongue in a chevron pattern as the water accelerates from the pool into the

Figure 1.8. The various features of a rapid are illustrated in this aerial photograph of Granite Rapid. *Photo Courtesy Glen Canyon Environmental Studies*

rapid. Lateral waves represent an intermediate stage in the water velocity between the subcritical (slow) flow of the pool area and the supercritical (very fast) flow of the rapid. At the bottom of the tongue, at the point where the water velocity passes from subcritical to supercritical flow, large *standing waves* form. The position of these standing waves relative to the shore changes at different flow levels. Scientists studying fluid dynamics refer to such waves as *hydraulic jumps.* At high flows these stationary waves can reach enormous proportions.

If the rate of flow through the rapid is not sufficiently fast to reach supercritical levels, standing waves cannot form. The waves that do form in this circumstance may actually migrate upstream or downstream. These *traveling waves* are generally small and of little consequence in the ride through a rapid. Near the bottom of the tongue of the rapid, lateral waves sometimes meet and form composite waves called *V-waves* or *haystacks* by river runners. These are characterized by explosive eruptions of spray at the crest. When such waves elongate, they are called *rooster-tails*. River running reaches a pinnacle of excitement in these waves, where a haystack can rise up and break just under a boat; more than one craft has slid off the side and capsized.

During the summers of 1984 and 1985 the Bureau of Reclamation's Glen Canyon Environmental Studies program sponsored a study of Grand Canyon rapids. Susan Kieffer, of the U.S. Geological Survey, was charged with the task of scientifically describing the repetitive pattern of pools, rapids, and high-velocity runouts. Her methodology included the unlikely and entertaining technique of releasing large inflatable models of the cartoon character Yogi Bear in the quiet pool above a rapid, as many as two or three at once, and timing their movement through the rapid. Notwithstanding the unorthodox appearance of her apparatus, Kieffer's work has led to a better understanding of how rapids actually perform.[10]

Her studies revealed a general pattern of continuing acceleration of water from the top of the rapid at the constriction of the channel, down the tongue, and continuing through the runout. The velocity finally begins to decrease in the diverging portion of the rapid (below the constriction) as the water spreads out in conformance with the widening channel. As the water begins to decelerate at the end of the runout, more nonbreaking lateral waves may form in the tailwaves.

The velocity of the river changes as a function of discharge (in this case, volume of water released from the dam) and the vertical cross section, or area of river. For example, water velocity generally ranges from 4 to 6 miles per hour (mph) in the straight runs between rapids, flows of 1 to 3 mph are common in the relatively quiet pools above rapids, and velocities over 20 mph have been measured in the steepest parts of a rapid. Though it often seems much longer, the average time a boat spends in the 100-yard-long, most violent section of Lava Falls is less than 15 seconds — an average speed of just over 20 mph. Though that is nowhere near the breakneck speed it seems to be to the boatman and his trusting passengers, a 20-mph impact with a canyon wall or partially submerged rock can cause significant damage to boats and bodies.

Although the configuration of the river channel at a rapid and the gradient of the riverbed are of great influence, the volume of Glen Canyon Dam discharges has at least equal control over hydraulic features of rapids. Standing waves illustrate this interrelationship. The standing wave in Crystal Rapid, for example, is made possible by the constriction of the river and the steepness of the bed, but the wave does not form across most of the navigable channel until dam discharges reach approximately 40,000 cfs. Discharges of this magnitude are well beyond power-plant releases (now a maximum of 31,500 cfs) and have occurred infrequently, but observations of Crystal Rapid at very high "flood" releases were possible during the summer of 1983. As the water rose, the rapid behavior changed markedly. At flow levels of about 60,000 cfs a memorable standing wave 15 to 20 feet in height formed an almost impenetrable wall across most of the river just above the narrowest part of the rapid. At higher releases the river forced a widening of the constriction, and the severity of the standing wave began to diminish.

Rapids can become more or less difficult to navigate as the water volume and velocity increase. Experienced river runners must sometimes choose between scheduling their runs at a low-water stage when they might be confronted with exposed rocks in the channel or, alternatively, running at a higher stage when the hydraulics associated with greater velocities might result in larger, more dangerous waves. Usually the higher water stages are preferred (Crystal Rapid at 40,000 to 60,000 cfs being the significant exception). The consequences of big waves are preferred to the dangers of exposed rocks and the damage they can do to motors, oars, inflatable pontoons, and fragile wooden boats.

Unless rock falls or tributary flows change the channel morphology from one trip to the next, experienced river runners know where the good runs are located. At flows less than 10,000 cfs, Crystal has a safe channel down the tongue and to the left of the hole where the standing wave appears at higher water; over about 10,000 cfs, the safe channel is to the extreme right, as far away from the standing wave as possible. In this case, the wise river runner would rather crash against the rocks on the right bank of the river than have to deal with the boat-flipping power of Crystal's standing wave.

Eddies, Whirlpools, and Boils

Most beginning river runners are startled when they first find themselves caught in a current that is carrying them (often rapidly and always inexplicably) upstream. They have discovered eddies, water that is spinning in the opposite direction of the main current. Eddies play an integral role in the formation of the sand deposits used as camping beaches, and from a boating perspective, they must be understood and continually dealt with if the river is to be successfully navigated.

Although eddy currents are strongest and most easily comprehended when associated with rapids, they can, and do, occur throughout the river. Any structure in the current, such as an underwater boulder, a bend in the channel, or a ridge of rock jutting out from shore, which increases the velocity of flow relative to the surrounding water or deflects the flow will cause an eddy to form. The development of eddy currents at the tail of a rapid is most typical. Eddies often form on either side of the high-velocity jet of water in the rapid runout, so that strong currents are flowing upstream along the banks on both sides of the river.

Eddies form below rapids because the widening of the channel below that point is usually abrupt. The river below the rapid can be as much as two to three times wider than it is at the constriction. The jet of high-velocity water escaping from the rapid shoots out so fast relative to the surrounding water that it maintains its constricted width far into the slower and wider portion of the river below. It is in the channel margin between the midstream, high-velocity jet and the riverbank that eddies form. As the jet flow moves downstream into the expanding, wider portion of the channel, it pulls along some of the adjacent, slower-moving water. This action sets the slower water to spinning in the opposite direction, much as one gear causes an interlocking gear to rotate the opposite way. The shear zone between the high-velocity stream moving in one direction and the slower water moving in the other is aptly called an *eddy fence*. Eddy fences may form a genuine, and sometimes formidable, boundary.

Because most of the camping beaches in the canyon are located just below rapids, river runners frequently are faced with the need to cross eddy fences to get to shore. Without a motor, it can be difficult to cross near the base of the rapids where eddy fences are always strongest. The strength of eddy currents dissipates with distance downriver, so boatmen usually wait as long as possible before trying to cross them. It is always a surprise to the inexperienced passenger to be carried hundreds of yards past the appointed campsite for the night, only to have the boatman enter the eddy at its farthest downstream point and then effortlessly drift back upstream to camp. It can be especially confusing if one boat is in the main current, and another is in the eddy — boats side by side floating in opposite directions in defiance of all apparent reason.

Whirlpools and boils often form in this zone of shearing currents. The development of really large whirlpools requires higher flows than normal dam discharges now allow, but the flood of June 1983 and exceptionally high water during the summers of 1984, 1985, and 1986 provided several memorable examples of these hydraulic features. In one instance during the summer of 1984, an 18-foot inflatable raft was flipped on a flat stretch of river by the combined action of a large boil on one side and a deep whirlpool on the other.

Waterfalls and Ancient Lava Flows

Dozens of large rapids punctuate the hundreds of miles of river in Grand Canyon, yet none is a waterfall in the classic sense. The specter of unavoidable and life-threatening waterfalls weighed heavily on the mind of Major Powell as he coped with the huge rapids of Cataract Canyon, more than 200 river miles above Lees Ferry, in late July 1869:

> there are great descents yet to be made, but, if they are distributed in rapids and short falls, as they have been heretofore, we will be able to overcome them. But, may be, we shall come to a fall in these cañons which we cannot pass, where the walls rise from the water's edge, so that we cannot land, and where the water is so swift that we cannot return. Such places have been found, except that the falls were not so great but that we could run them with safety. How will it be in the future?[11]

We now know that there are no waterfalls on the river where differential erosion of bedrock has produced a resistant ledge. The well-known "ledge hole" in Lava Falls Rapid was thought for many years to be caused by a resistant basalt outcrop in the channel, but it is actually due to a row of several large basalt boulders brought in by a prehistoric debris flow down Prospect Canyon.

No waterfalls exist in the canyon today, but they may have occurred in the distant past, especially where ancient lava flows blocked the river. The most recent such flow, the Toroweap Cascade, spilled over the canyon rim approximately 140,000 years ago very near the site of Lava Falls. Evidence suggests that the resulting dam was more than 225 feet high, creating a lake in the canyon at least 40 miles long.[12]

The River's Burden

In the millennia before man-made dams on the Colorado River, the river indefatigably pushed onward to its mouth at the Gulf of California. There its water nourished a productive estuary and delivered sediment to an ever-growing delta. These features began to decline, and later to disappear, after Hoover Dam and other water-diversion projects interrupted the natural flow of nutrient-rich water and silt.

Water was only one of the two treasures that the Colorado River carried to nourish its estuary, the other being the silt, sand, and gravel eroded from deserts and mountains throughout the entire river basin. These sediments continued to be carried through Grand Canyon until Glen Canyon Dam was completed, at which time the river's natural burden became trapped at the bottom of Lake Powell. The normal downstream movement of erosional debris has now been interrupted. We are only beginning to understand and, to the extent possible, accommodate these and the other consequences of humans' manipulation of the once-wild river.

2

SAND AND ROCK
Sediment and Substrate

Crystal Rapid: the mere name sends a tremble of apprehension through anyone who must pilot a boat past the standing wave and merciless hole of one of the West's most notorious rapids. Strangely enough, Crystal Rapid was not much more than a pleasant riffle before late autumn of 1966. The rapid had a modest drop, and the few large boulders in the channel could easily be avoided regardless of the water level. Remote, inaccessible, and unseen except by the few rafters who ran the river in those days, Crystal was tiny in comparison to the larger rapids just upstream: Horn Creek, Granite, and Hermit. Events taking place over the nearby Kaibab Plateau in early December 1966, however, instantly changed Crystal into one of the most difficult rapids in Grand Canyon.

An intense, unseasonal storm from the Pacific Ocean moved slowly into northern Arizona during the night of December 3. Over the next three days, at least 14 inches of relatively warm winter rain fell on the north rim in an area centered at the head of the Crystal Creek drainage. Fourteen inches of rain in an entire year is a considerable amount for most of this region; to receive it in just a few days is extraordinary. The Crystal drainage has not experienced a comparable deluge in recorded history.

The flood that followed the storm was what hydrologists refer to as a *probable five-hundred-year event*. To add to the severity of the subsequent runoff, approximately 6 inches of snow lay on the ground when the rain began to fall. Melted by the warm rain, the snowpack contributed its waters to the gathering flood. First the water trickled downslope in a myriad of rivulets, with each small drainage joining a larger one, adding to the flow until every headwater of the drainage ran a torrent. The floodwaters, swelling all the while, raced over the canyon rim to join Crystal Creek in its 13-mile tumble to the Colorado River 6,500 feet below.

Thunder at Crystal Creek

Plunging down the already waterlogged and unstable slopes, the flood set off widespread landslides in the soft Hermit Shale and Supai formations. Tons of loose rock and soil mixed with the water to form what geologists term a *debris flow*. Consisting of boulders, gravel, sand, silt, and water bound in a slurry, a debris flow is a powerful erosive agent. The flow rumbling down Crystal Creek in early December 1966 was a large one. Estimated at a depth of 44 feet, with a discharge exceeding 10,000 cfs, the flow ripped out large cottonwood trees, inundated Anasazi Indian ruins that had stood unscathed for eight hundred years, and deeply eroded the stream channel.[1]

Boulders measuring 5 feet in diameter, picked up and carried along like so many bobbing corks, were common in the debris flow. Even boulders more than 14 feet across and weighing nearly 50 tons were moved downstream, sometimes great distances.[2] The momentum gathered during the flow's journey from rim to river delivered hundreds of tons of boulders, sand, and gravel far out into the Colorado, with some debris probably reaching the opposite shore. By the time the chaos had quieted, a deep layer of rock and sediment had been injected into Crystal Rapid, effectively damming most of the river for a few moments. The impounded river swiftly rose behind "Crystal Dam," then flowed over the top and cut a new channel down the left (south) side. Crystal Rapid would never be the same.

The peak of the debris flow lasted only a short time, probably less than 30 minutes, but in that time Crystal Rapid was transformed from a mild riffle into a serious barrier to safe river navigation. Before the debris flow the width of the channel in the rapid had been about 280 feet. After the flow the boulder-strewn channel had narrowed to 100 feet, and a rock pile in the path of the runout greatly increased the difficulty of successfully maneuvering the rapid.

Debris Flows: A Canyon Phenomenon

A landscape like Grand Canyon almost demands the occurrence of debris flows. High relief, steep canyon walls, an abundance of unstable talus slopes, and a variety of rock layers eroding at different rates all combine to create a high potential for landslides. Torrential rainstorms can easily trigger slope failures, the loose material joining the runoff to flow downhill until the gradient becomes too shallow, or the water content drops too

Figure 2.1. Crystal Rapid before (*top*) and after (*bottom*) the debris flow of December 1966. The upper photograph was taken in May 1966 at a riverflow of approximately 16,000 cfs; the debris fan is very small, and a smooth tongue of water enters the rapid just to the right of the center of the river. The lower photograph was taken in April 1986 at 28,500 cfs. Notice how the tongue of the rapid has been moved substantially to the left by the debris fan deposited during the debris flow of December 1966. *Top photo by J. Harvey Butchart; bottom photo by Raymond M. Turner*

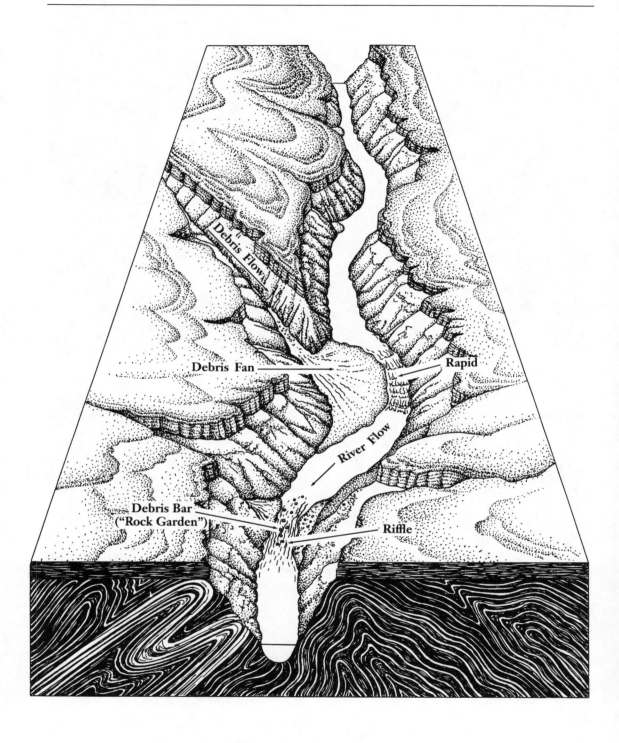

Debris Flow

Debris Fan

Rapid

River Flow

Debris Bar
("Rock Garden")

Riffle

low, to support further movement. If the flow reaches the river, it can deposit sufficient boulders and other sediment to constrict the channel and force it against the opposite bank. With the power of these flows, even small drainages can exert substantial control over the hydraulics of the river. Recent examination of fifty-seven major rapids in Grand Canyon has revealed that fifty-four were caused by debris flows.[3]

Containing 10 to 40 percent water by volume, debris flows are the low-water extreme of three types of streamflow classified on the basis of water content. All three flow types have been casually labeled flash floods. Normal streamflow is composed of 80 to 100 percent water, and hyperconcentrated flow, which looks and acts like normal flow, contains 40 to 80 percent water. The low water content of debris flows can make them appear more solid than liquid, often with a consistency resembling that of flowing concrete. The main body of the material travels as a large plug, or peak. A debris flow event may be brief, involving only one peak, or may extend over a period of minutes to hours with several peaks separated by inactivity.

Even with a water content under 20 percent by weight, debris flows can move deceptively fast. They are known to travel at speeds ranging from 10 to 18 feet per second (7 to 12 mph), about as fast as a terrified person can attempt escape.

Robert Webb, an earth scientist with the U.S. Geological Survey, studied debris flows in Grand Canyon as part of the Glen Canyon Environmental Studies. His investigations revealed that debris flows are the primary method of sediment transport in small, steep drainages to the Colorado River. The average annual input of sediment to the river by this means is unknown, but it must measure in the hundreds or even thousands of tons. The combined input from all debris flows to the river in some years may exceed one hundred thousand tons, depending on the frequency and duration of rainfall.

An average of one or two debris flows appears to reach the river each year. In periods with unusually intense rainstorms, multiple flows may hit the Colorado as the result of a single storm. That is exactly what happened during the summer of 1983 at the mouths of almost all the tributaries from River Mile 42 to River Mile 77. These flows probably occurred during the thunderstorm of July 25.[4]

For every debris flow that reaches the river, many more begin but fail to cover the full distance. The flows that have stalled out leave deposits of sediment that remain like traps waiting to be sprung. The right thunderstorm the next summer or even,

Figure 2.2. Debris flows from tributaries form most Grand Canyon rapids by constricting the channel and forcing the current to the opposite side of the river. Other river-channel features, such as pools above rapids and rock gardens below rapids, are also controlled and maintained by debris flows.

perhaps, the next century, can transform the waiting plug into an active flow in a matter of moments.

The frequency of debris flows varies from drainage to drainage. Those tributaries exhibiting the most flows generally have larger rapids at their confluence with the river. Debris flows have reached the river at least three times in the Lava-Chuar Creek drainage during this century: once between 1916 and 1966, once in December 1966 as a result of the same storm that enlarged Crystal Rapid, and once between 1973 and 1984, for an average of one every twenty to thirty years.[5] Crystal Creek drainage has averaged a minimum of one flow every fifty years over the last several hundred years, although only one has reached the river so far in this century. Monument Creek, a tiny drainage of only slightly more than 3 square miles, has had at least two major debris flows this century. The first one occurred about 1960, as indicated by growth-ring counts of catclaw acacia trees scarred during the event. The more recent debris flow happened on July 27, 1984, when a thunderstorm triggered a flow of about 4,000 cfs. This flow reached the river, where it greatly modified Granite Rapid at the mouth of Monument Creek. Boulders weighing thirty-seven tons were transported by this flow.

Few people have ever seen a debris flow, and fewer still have had to run from one, but those who have describe the sound as being like a dozen locomotives approaching at top speed. One Grand Canyon river guide who knows the terror of being caught in the path of a large flow is Mike Walker of OARS, a commercial river-running company. His hair-raising experience took place in mid-July 1984.

Tired from de-rigging boats at the end of a twelve-day Grand Canyon adventure, and ready to return to the reality of the outside world, Mike's group had just boarded its trucks for the bone-jarring ride up the primitive road that runs from the mouth of Diamond Creek to Peach Springs, Arizona. Mike was driving one of two large trucks loaded with people and river-running gear. He was no more than a half-mile from the river, maneuvering his way up the streambed that passes for a road in the narrowest part of Diamond Creek Canyon, when his attention was caught by an unusual noise. What he heard was the sound of a 7-feet-high wall of earth and water surging down the canyon toward him.

Jumping from the trucks and dashing to safety, Mike and his companions barely beat the flood. Gaining high ground, they looked back in time to witness the fate of their abandoned vehicles. Both trucks, loaded with thousands of dollars of river gear, fourteen boats in all, were lifted by the torrent as though made of balsa wood and then swept downstream, around a bend, and out of sight. The occasion was not without its black humor. Stunned into what had to be total disbelief, and no doubt still shaking from the experience, Mike realized that he had dutifully removed and pocketed the keys to the truck he had been driving. With a shrug, he pitched the now-useless keys into the flow after the trucks and settled back to wait for rescue.

Farther downstream, eyewitnesses still unloading their boats at the mouth of Diamond Creek reported seeing a wall of spray 30 feet high when the flow hit the Colorado River. The flash flood was sustained for six and one-half hours. It carried tons of boulders and sediment into the river, greatly enlarging the alluvial fan at the mouth of Diamond Creek. Remarkably, as crowded as Diamond Creek was that day, not a single life was lost, nor was there even significant bodily injury. The scant remains of the mangled trucks were later discovered in the Colorado River, more than 1,000 feet downstream of the Diamond Creek confluence. For weeks, pieces of the wreckage surfaced along the river, little of it salvagable.

The Changing Substrate

The dramatic, overnight transformation of Crystal Rapid and the close call at Diamond Creek serve to remind us that the Colorado River corridor is a dynamic and ever-changing system. Alterations can be almost instantaneous, such as those caused by debris flows, or more characteristically, prolonged, as in the system's response to the river's cycle of rise and fall.

A key element in this system, second only to the life-giving water itself, is sediment — sediment that hangs suspended in the current, lines the riverbed, and lies exposed in beaches and terraces along the shore. This sediment, in its many sizes from clay particles to boulders, shapes the river's channel and forms the substrate that ultimately influences most life forms in the canyon corridor.

For millions of years the river and its tributaries have cycled sediment through the canyon in a complex pattern of erosion and deposition. At any given moment the river is both picking up sediment and dropping it, storing it and moving it, depending on the current flow in any particular spot. This pattern shifts constantly in response to changing river discharges and velocities.

Sediment in Storage

Deposits of riverborne sediments, consisting of the finest clay
particles, grains of sand, gravel, cobbles, and even boulders, are
all called *fluvial deposits*. The entire river channel — bed, banks,
and high terraces — is formed by the action of flowing water
dropping its sediment load as water velocity slows. In every case
the beaches on which we camp and wildlife flourishes originated
under water. Deposited on the river bottom as sandbars during
high flows, these sediments were left exposed as the water level
dropped. The lowest deposits, nearest the shore, are often
resubmerged and reworked on a daily basis as the river rises and
falls. The highest deposits, however, are now beyond the river's
reach. Marking the extent of the great floods of pre-dam days,
these deposits stand as isolated terraces high above the present
river level.

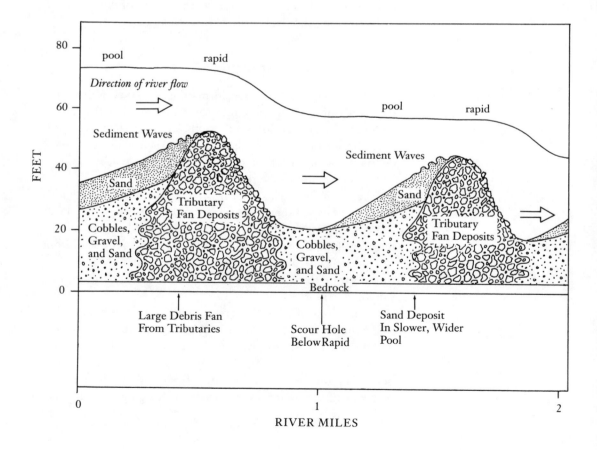

All riverbank and river-bottom deposits fall into three broad categories based on particle size. Sediment is sorted by the current into distinct deposits that generally contain particles of predominantly one size.

Alluvial fan deposits at tributary junctions comprise the category with the largest particle size. Often called *boulder* or *debris fans*, these debris-flow deposits are originally composed of different-sized sediment, but the smaller particles eventually wash away, leaving only boulders. The large size of the boulders makes them relatively immobile.

The second major type of sediment accumulation is the *cobble* or *gravel bar deposit*. Cobble bars are made up of well-rounded rocks fewer than 18 inches in diameter. They develop in the river in places where the current loses velocity and can no longer move sediment of this size. Cobble bars often appear where the river abruptly increases in width or as a mantle over boulder fans on the inside bend of rapids or riffles. Cobbles generally cannot be transported by the controlled flows from Glen Canyon Dam, so most present-day cobble bars are relicts of natural floods that swept the canyon before 1963.

Sandbars are the third type of fluvial deposit. Sand deposits make up the majority of the riverbed except in rapids and riffles with swift current. They also form the beaches that commonly define the riverbanks, except in the Granite and Muav gorges. These are the beaches so crucial to streamside vegetation and the animals it supports — to say nothing of the thousands of river runners who camp along the river every year.

Because of their small size, sand grains are one of the few kinds of movable sediment that controlled flows from the dam are able to transport at most discharges. Sand that is not being actively transported by the river is stored either on the bottom of the channel or in eddies. Sand stored in eddies is usually very well sorted and fine- to very fine-grained in size, whereas sand found on the bed of the channel is medium-grained. This pattern of deposition is amazingly consistent.

The sand deposits along the river in Grand Canyon have been studied and interpreted by Jack Schmidt and Julia Graf, geologists specializing in the origin of landforms created by rivers. After spending months digging holes and trenches in sandbars to examine the characteristics and development of sand deposits, Schmidt, Graf, and their co-workers identified four distinct types of deposits.[6]

Two of the four types develop within eddies. Separation deposits cover the downstream edge of boulder fans at tributary

Figure 2.3. A depth profile of the river shows the continuous sequence of deep pools below shallow rapids. Patterns of riverbed material follow the same pool-drop sequence.

junctions. They form near the point where the eddy flow separates from the main current, where the eddy begins. Reattachment deposits are located at the downstream end of eddies and project upstream into the eddy center. They develop near the point where main-channel flow and eddy flow are reunited (or reattached) to form an uninterrupted downstream flow. Separation and reattachment deposits are the largest and most numerous types of sandbars. They normally rise the highest out of the river and provide excellent campsites for river runners. For example, the main camp at Saddle Canyon is on a separation deposit, and the lower camp at Saddle Canyon is located on a reattachment deposit.

The current in an eddy flows upstream in a circular way. Sediment-laden water from the main channel slows as it enters the eddy, dropping its suspended load of sand to form the reattachment bar. The water then picks up speed as it circles upstream in the eddy, gaining enough velocity to cut a small channel through the newly deposited reattachment bar as it returns to the main channel. This cut is called the *return current channel*. A secondary eddy, located upstream of the primary eddy, receives sediment-laden water from both the return current channel and the main river channel. The secondary eddy has a lower current speed than either of its feeder sources, so it drops sand out of suspension to form the separation bar.

All the sand stored in an eddy may be scoured out by peak flood flows. However, reattachment and separation bars similar to the ones present before the flood will begin to form underwater as soon as the crest of the flood starts to recede. The new sand deposits will not be identical to the previous landform if the shape of the boulder fan forming the eddy changed in any way, if the river stage fell faster or slower than it did during the formation of the previous sand deposits, or if the river was transporting a different kind of sediment. Reattachment bars are more susceptible to erosion by fluctuating river levels than are the more stable separation deposits.

The two other types of sand bars are *upper-pool deposits* and *channel-margin deposits*, neither of which is associated with eddies. Upper-pool deposits cover the upstream edge of boulder fans at the head of constrictions forming rapids. Swift, sediment-laden water in the main current temporarily "pools up" and loses speed near the boulder fan above a rapid, causing the sand to drop from suspension to form the upper-pool deposit. Channel-margin deposits, as their name implies, are located along the riverbank away from rapids and eddies. They are created where water in the main channel slows and drops its sediment load as it

Figure 2.4. Patterns of riverflow (*top*) and the resulting shape of riverbed sand deposits (*bottom*) are consistent throughout the river corridor near debris fans forming rapids.

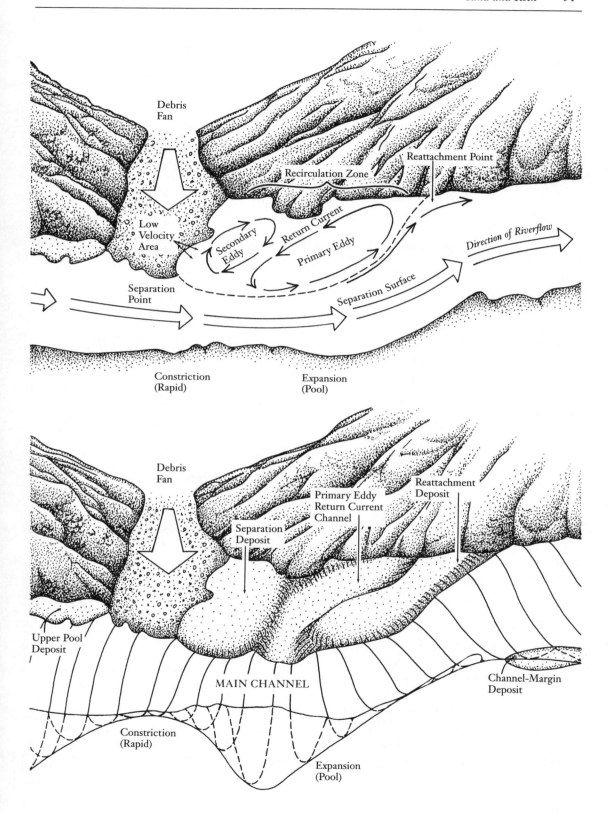

Debris
Fan

Reattachment Point

Recirculation Zone

Return Current

Low
Velocity
Area

Secondary
Eddy

Primary Eddy

Direction of *Riverflow*

Separation
Point

Separation Surface

Constriction
(Rapid)

Expansion
(Pool)

Debris
Fan

Primary Eddy
Return Current
Channel

Reattachment
Deposit

Separation
Deposit

Upper Pool
Deposit

Channel-Margin
Deposit

MAIN CHANNEL

Constriction
(Rapid)

Expansion
(Pool)

encounters the friction of the riverbank, forming long, narrow sandbars.

Old versus New Sediment Transport

The construction of Glen Canyon Dam did not stop this process of sediment transport, erosion, and deposition; the process still goes on, but in a drastically altered way. Before the dam was built, the Colorado River carried uncounted billions of tons of the eroding land mass through the Grand Canyon. The amount of sediment borne by pre-dam floods was staggering. On an average day before 1963 about 380,000 tons of sediment was transported past Phantom Ranch. This amount is five times greater than the weight of the H.M.S. *Titanic*, one of the largest ocean-going vessels ever built. In one record-breaking day more than 27 million tons of sediment was carried past Phantom Ranch at the peak of the great flood of 1927.

After construction of Hoover Dam, all sediment transported through Grand Canyon eventually ended up in the reservoir of Lake Mead. Sediment-laden river water encountered the still water of upper Lake Mead and dropped its load to form large beds of sand and silt. These silt beds continued to grow until the construction of Glen Canyon Dam in 1963, pushing the delta of the Colorado River farther and farther downstream toward Pearce Ferry. By 1949, the accumulation of sediments was more than 270 feet thick in places, obscuring the original river channel, raising the river level, and decreasing the water-storage capacity of Lake Mead by about 5 percent.[7]

Some of the most notorious rapids of the lower Granite Gorge, including Separation and Lava Cliff rapids (which had caused such difficulty for Major Powell in 1869), were covered by deep layers of sand and silt. Sediment deposited in the river channel between Bridge Canyon and Pearce Ferry reshaped the river gradient from its original 8 feet per mile to a gentler slope of only 1 foot per mile. Here the Colorado was transformed into a stream with mature characteristics more resembling the Mississippi River than the vigorous, young river that still flowed through the Grand Canyon upstream.

With the construction of Glen Canyon Dam, however, this massive load of sediment began to be trapped farther upstream in Lake Powell. The river entering Grand Canyon now must pass through the dam. It is clear and entirely free of sediment, with enormous potential to erode but little ability to deposit. As a result, all the new sediment needed to replenish eroded

Figure 2.5. The Colorado River deposited 2 billion tons of sediment at the head of Lake Mead in the lower Granite Gorge after the completion of Hoover Dam in 1935, forming extensive silt beds and mud flats that are exposed at low water. This 1949 photograph shows craters, measuring approximately 7 feet across and 4 feet deep, formed by the release of methane gas generated by decaying organic materials in the silt beds. *W. O. Smith (#8), Courtesy U.S. Geological Survey Photo Library, Denver*

Figure 2.6. Sediment deposited by the Colorado River at the head of Lake Mead from the 1930s to 1963 formed an ever-advancing delta that covered the original riverbed with sand and silt to a depth of hundreds of feet.

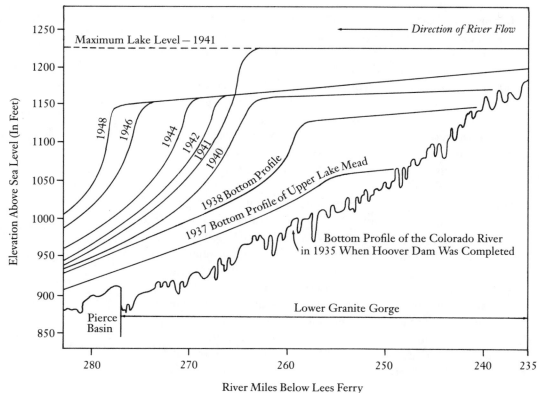

deposits in the river corridor must come from tributaries within the canyon.

The average annual tributary input of sediment has been estimated at almost four million tons, with the Little Colorado and Paria rivers contributing about three-fourths of the total. Rare tributary floods may briefly increase river flow by as much as 10,000 cfs, depositing sand and silt terraces up to 18 inches thick overnight. A single debris flow may bring in more than 300,000 cubic feet of sediment, one-quarter of which is sand. In spite of these impressive figures, the Colorado River's present sediment load is only a fraction of pre-dam amounts. On the average, only 40,000 tons of sediment is now transported past Phantom Ranch each day, less than one-ninth the pre-dam figure.

Before the dam, spring and summer floods swept the canyon in a natural cycle of scour and fill that annually renewed and replaced the banks and bed of the river. Cold-water spring floods caused by melting snow in the Rocky Mountains reached their peak in June, transporting primarily coarse sand. Warm-water summer floods resulting from thunderstorm activity reached their peak in August or September. They transported primarily silt and clay.

Because the sediment-transport capacity of water increases at lower temperatures as water molecules become denser, flowing water grows more erosive as it becomes colder. As a result, pre-dam spring floods eroded the riverbed, commonly causing a net scour of 20 feet or more. In contrast, summer floods were less erosive, usually leaving a net deposition of sediment roughly equivalent to the net loss caused by spring floods. The great depth to which scour and fill could occur under pre-dam conditions was revealed at Hoover Dam in the early 1930s by workers who discovered a saw-cut timber buried under 50 feet of sediment.[8]

Glen Canyon Dam now prevents annual flooding, altering this natural cycle of scour and fill. The clear, cold water released from the dam does not flow passively through the canyon. Far below its potential capacity to transport sediment, the water scours sediment from the riverbed, picking it up and carrying it downstream. River runners have observed that the water that was clear and bright green at Lees Ferry is usually clouded with a small amount of sediment by the time it gets to Diamond Creek, even in June when there are no thunderstorms to make tributaries run muddy. Recent studies have clearly demonstrated that the beach and bottom sediments of the river are leaving the system, washing downstream to accumulate in the giant delta under the water of Lake Mead.

Not Enough Sand, Too Many Boulders

Paradoxically, at the same time that the river's increased capacity to remove fine sediments from the system is causing one problem, its reduced capacity to remove the largest sediment is causing another. Because the dam now prevents the massive scouring floods in excess of 100,000 cfs needed to move larger rocks and boulders brought in by debris flows, these sediments remain at tributary mouths to encroach more and more on the river each year. In the first ten years after Glen Canyon Dam was completed, 27 percent of the alluvial fans at tributary mouths had built outward because of debris flows; 10 percent of these fans expanded by more than 50 feet.[9]

If the controlled discharges from the dam are inadequate to move the largest of the boulders placed in rapids by debris flows, rapids will become increasingly difficult to run. The post-dam debris flow at Crystal Creek and what it did to Crystal Rapid graphically illustrates this probability.

One measure that can reflect a rapid's severity is its constriction ratio: the ratio of the width of the narrowest portion of a rapid to the average width of the channel upstream of the rapid. Geologists calculating the constriction ratios of Grand Canyon rapids have found that most are at or near a ratio of 0.5, meaning that the river is one-half as wide at the rapids as elsewhere. This consistent measurement indicates that the river has eroded rapids with remarkable uniformity over time.[10] In contrast, the constriction ratio of Crystal Rapid after its transformation in 1966 was a narrow 0.25. This low ratio is probably characteristic of recently formed Grand Canyon rapids with steep gradients, rapids that approach the limits of navigability.

The flood of June 1983, cresting at more than 92,000 cfs, rearranged Crystal Rapid and increased the constriction ratio to 0.35, but a flow of at least 400,000 cfs would be required to enlarge Crystal to the standard 0.5 constriction ratio. Such a flow is not going to happen in the foreseeable future. Crystal Rapid will be with us for a long time as one of the most severe rapids in Grand Canyon, and more debris flows of the magnitude that intensified Crystal are bound to occur in other drainages. It is conceivable that the next one-thousand-year storm, due in a few centuries or at the end of next week, could

set loose a river-constricting debris flow that would all but
preempt safe boating through Grand Canyon.

Sand in Motion

The 15-mile stretch of river in Glen Canyon located between
the dam and Lees Ferry was first to show the effects of sediment
loss caused by the dam. Erosion and loss of sediment is not
constant from the dam to Lake Mead, but instead follows a
pattern of decreased erosion with increasing distance from the
dam. The riverbed in Glen Canyon was scoured 27 feet in the
first few years after 1963, largely because of the unusual release
of 40,000 to 60,000 cfs for forty days in 1965.[11] Fine-grained
sediments were washed downstream during the high flows,
leaving only those sediments too heavy and coarse-grained to be
moved. The riverbed in Glen Canyon is now largely armored
with gravel, cobbles, and boulders and is not expected to erode
further.[12]

The substrate of the river from Lees Ferry to Phantom Ranch
has responded similarly to the effects of controlled flows from
the dam, but on a smaller scale. Riverbed deposits throughout
the river corridor were initially eroded during the first few years
after the dam began operation but quickly recovered because of
sediment provided by tributary flow. The riverbed has an
extraordinary ability to store sand, as indicated by a net riverbed
fill of 6 to 7 feet by 1970.[13] A portion of this new sediment
evidently came from the riverbanks that were exposed by low,
fluctuating flows, as many camping beaches experienced lateral
erosion averaging almost 3 feet per year in the 1970s.[14] The
increase in riverbed sediments was taking place at the expense of
the beaches. In essence, silt and sand were carried by the current
from the beaches out into the main channel, where they slowly
accumulated on the river bottom in low-velocity pools.

The flood of 1983 further eroded the riverbed between Lees
Ferry and Phantom Ranch, removing an estimated 16 million
tons of sand and other sediment from the channel. Subsequent
fill of the riverbed was rapid after the flood, and riverbed
equilibrium was reached in approximately six years. The peak
discharge of the 1983 flood, about 92,000 cfs, was responsible
for most of the scour and sediment loss. If the peak of the flood
had been magically restricted to 28,000 cfs, but with the same
volume of water released over a period of weeks or months, the
volume of sediment lost and the resulting time needed to
reestablish equilibrium would have been reduced by one-third.[15]

Sand deposits downstream of Phantom Ranch are largely intact, with some exceptions, because the river's sediment balance more closely approaches equilibrium with increasing distance downstream from Glen Canyon Dam. Sediment contributed by tributaries increases with distance below the dam, and riverbed scour decreases. The input of sediment and the ability of the river to transport that sediment are almost equal below Phantom Ranch. As a result, there was little if any net loss of sand from the riverbed in that stretch during the 1983 flood.

Beach Campsites — A Disappearing Resource

Every Colorado River boatman with at least ten years of experience emphatically insists that most campible beaches have decreased in size through time. Boatmen can point to camps that are no longer used because the sand has been washed away. Conversely, they can point to rocky, armored areas of other beaches that formerly were blanketed with deep layers of sand. They know when the changes occurred, too. A small but perceptible decrease in the beach area at the water-land interface was observed from the late 1960s to 1983.

Early attempts were made to quantify these anecdotal observations that the beaches were getting smaller. In an effort to determine the rates of erosion of sand from campsites, Alan Howard, of the University of Virginia, surveyed several beaches from 1974 to 1975 and established a series of semipermanent bench marks.[16] Many of these sites were resurveyed on a regular basis beginning in 1980 by Stanley Beus, of Northern Arizona University, and other researchers. The survey profiles provide a chronology of mostly deteriorating beaches.

Only slight changes in the beach profiles were recorded during the period from 1974 to 1982. Some beaches lost up to 3 vertical feet of sand, while others actually gained 1 to 2 feet. Overall, slightly more sand was lost than gained, suggesting a slow and gradual depletion of sand from the beaches studied.[17]

The sediment budget and resulting beach profiles changed dramatically after the 1983 flood. More changes in the beaches occurred during the few weeks of spillway releases than in the preceding eight years, and when the floodwaters subsided, most of the beaches were unrecognizable. One of the peculiarities of the flood was a tendency for major deposition on the upper beach terraces, while lateral erosion cut the lower face of the beach away. There may have been more sand stacked against the canyon walls, but beaches were reduced in size.

Other changes in the beach substrate were noted as well. Sand grain size was larger, and the percentages of organic composition and nutrient levels were lower.[18] The finer-grained silt and clay particles remained in suspension and passed on to Lake Mead, leaving behind nearly sterile deposits of pure, coarse-grained sand. Even the pre-dam beaches that remained were covered by a thin mantle of sterile sand. As a result, fine-grained organic particles high in nutrients were leached from the system. These changes benefited river runners, who value clean camping beaches, but were possibly detrimental to the long-term development of riparian vegetation.

Continued high water during the summer of 1984 removed much of the sand gained during the 1983 event. After 1984, continued beach surveys document a consistent pattern of sand and campsite loss. Of the original twenty beaches studied, by 1985, four were gone, eleven had significantly decreased in size, and only three had actually gained sand.

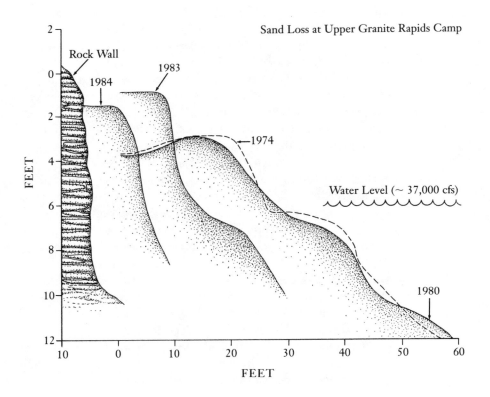

Trapped Sediment — Clear Water

There is little question that the sediment balance within Grand Canyon's river corridor is swinging toward a loss of sand. This loss could have lasting consequences for the controlled, sediment-limited river of the future. Ultimately, concerns about the substrate of the river must address the long-term balance of sediment. Will there be an additional net loss of sand over the next twenty, or fifty, or one hundred years?

The answer to this question is unknown but may well lie with how the dam is operated. For example, steady releases of a constant amount of water have the greatest potential to store sand, whereas fluctuating water levels have the least potential to store sand. Some earth scientists have predicted that a gradual loss of beaches is inevitable, especially where the river runs through narrow gorges, and that substantial losses may occur in the future.[19]

Such depletion is a grave concern because the sediments making up the physical substrate of the river are vital in providing habitat for riparian life, as well as campsites for river runners. But the situation is not one-sided. Paradoxically, even as the clear, erosive river threatens riparian life on the one hand, it promotes aquatic life on the other. For by removing the sediment from the river, Glen Canyon Dam has added a new dimension to the aquatic ecosystem. This new dimension is light, in the form of life-generating sunlight that is able to penetrate the now-clear river water to bring about profound changes. One of the most interesting aspects of the story of the new river is that the dam-caused alterations in the physical environment of the Colorado River have carried over into the other realm of the river ecosystem — that of the living organisms — producing far-reaching changes for virtually all plants and animals downstream of the dam.

Figure 2.7. Some camping beaches in Granite Gorge were eroded a horizontal distance of more than 30 feet from 1974 to 1984, as at upper Granite Rapids camp.

PART II   Life in a Desert River:
            The Aquatic Ecosystem

3

BUILDING BLOCKS
Primary Producers and Aquatic Invertebrates

On the evening of March 13, 1983, Robert Jensen, Colorado River whitewater guide and fishing enthusiast, was drift-fishing 7 miles below Glen Canyon Dam. Jensen was pursuing the same quarry the other one hundred or so fishermen on the river that day were after: a trophy rainbow trout. And this was the place to find the really big ones. March 1983 was two decades to the month after Glen Canyon Dam unseated Mother Nature as the master of the Colorado River in Grand Canyon. In that time the stretch of river from the dam to Lees Ferry had developed into a phenomenal trout fishery, attracting anglers by the thousands.

That evening will be remembered by Jensen for a long time, for it was finally his turn to catch a rainbow trout in excess of ten pounds. He was dragging more than two hundred feet of almost invisible four-pound test monofilament line, a necessity because the fish would be able to see anything heavier in the clear water. At the end of his line was a tiny weighted hook, covered with pinkish-green, tightly woven yarn, expertly wrapped to look like a miniature shrimp.

This fishing technique required a certain knowledge of the fish habitat and habits. Jensen had been drifting over the same quarter-mile section of river for at least an hour. His fishing spot covered dense beds of large, filamentous algae called *Cladophora*, with thousands of individual plants blanketing the bottom of the river. Although his hook constantly hung up in the algae and sometimes temporarily snagged on the bottom gravels, he knew this habitat was a favorite feeding area for the big trout. In spite of the frustrations, he persisted. The water was so clear that occasional glimpses of large silhouettes gliding beneath the algae was evidence enough that the big ones were close. The trout were after tiny crustaceans (amphipods), the superficially shrimplike creatures Jensen's lure was meant to imitate.

Trophy Trout in the "New" River

He still tells the story with an edge of excitement that could
only be a whisper of the heart-pounding tension he must have
felt during the actual contest:

> I was about ready to call it a day. For the past several hours I had
> had only a couple of bites, nothing much was happening, and I
> wanted to get back to camp before dark. Suddenly my line went
> tight, and I instinctively twitched the rod to set the hook, only to
> feel the drag typical of a large glob of algae. I began reeling in the
> line. It felt really heavy, but not the kind of heavy fighting that
> indicates a fish. When my line was about 30 feet from the boat, I
> could see a patch of green algae on the end of it, and what looked
> like a big, thick stick. Then, the stick began to move — it was the
> biggest rainbow I had ever seen. The fight was on. Six times in
> twenty minutes I had the monster up to the boat, and each time
> but the last he dashed back into the deep water, stripping my thin
> line as he dove. Several times he broke water in great
> lunges . . . never have I been more excited.

Lees Ferry inspires literally hundreds of fishing stories just like
Jensen's, and a large proportion are actually true! But even more
amazing than the size and quantity of the fish is the fact that this
trophy fishery appeared in a river that was incapable of
supporting a game species like trout before the dam was built.

The ecological story and sequence of events leading to the
development of such large rainbow trout can still be read by the
observant biologist in the gut contents of these fish. Their
stomachs are packed with bright green algae, midges, black flies,
and amphipods. On closer examination the green algae are
found to be covered with millions of microscopic but energy-
rich diatoms. This highly nutritious diet could not have
developed in the muddy desert river that flowed through Glen
and Grand canyons before the dam was built. It is typical,
however, of the stream the Colorado has become. The trout's
stomach contents reflects a great increase in aquatic productivity
that is a direct consequence of Glen Canyon Dam.

Because the dam traps the muddy water of the river in Lake
Powell and releases clear water in its place, sunlight, the most
important source of energy for the initiation of plant and animal
growth, can now penetrate the stream. This clear water is the
single most significant factor in the development of a radically
changed river and aquatic ecosystem. And as we shall see, it has
had far-reaching consequences for all living elements of the
Grand Canyon river environment.

Figure 3.1. Trophy rainbow trout, such as this one caught above Lees Ferry, are a direct result of enhanced aquatic productivity due to the clear water released from Glen Canyon Dam. *Steven W. Carothers*

Productivity in Aquatic Environments

One of the most obvious results of the newly available sunlight has been a tremendous increase in the growth of algae. Since no pre-dam information was gathered on productivity rates and distribution of algae, it is not possible quantitatively to report the increase in biomass produced when the river changed from muddy to clear. Algal beds now providing the biological foundation for the river's food web simply did not exist in the pre-dam river, for they could not have grown in the turbid water.

The amount of solar radiation penetrating the water column as a result of the new water clarity is critical in determining post-dam aquatic productivity. Other physical and chemical factors, however, including water temperature, nutrient levels, frequency of water fluctuations, bank and bottom conditions, substrate stability, and water velocity, mediate plant and animal growth within the river ecosystem. Each of the conditions influencing productivity can be altered, sometimes radically, by simple changes in how and when water is released from Glen Canyon Dam. Thus the manner in which the dam is operated can make a significant difference in the quantity and quality of food-web productivity.

Primary productivity is the very beginning of food-web development. The process starts with what are called *autotrophic organisms* — life forms (in aquatic ecosystems, usually a variety of algal species) that can make their own food from inorganic materials. Although certain types of biochemically induced primary productivity can take place in the absence of light, solar energy is responsible for virtually all stream productivity. Sunlight empowers green plants to manufacture their own food through the biochemical process of photosynthesis, literally "creating from light." In any body of water, whether lake or river, water serves as a medium within which green plants use chlorophyll — a complex molecule similar to the hemoglobin molecule that transports life-giving oxygen in the blood of higher animals — to convert carbon dioxide and water into organic substances. Simply put, green plants are able to "feed on light" because of chlorophyll. The food web of the river corridor in Grand Canyon originates in the depths of the river, where single-celled algae and more complex attached algae flourish.

Physical Changes Influencing Aquatic Growth

Light Transmittance

Measurements of light penetration into the pre-dam Colorado River are not available, but comparisons of the post-dam river and the Little Colorado River in flood have been made. Under flood conditions the Little Colorado is comparable to the undammed Colorado. Light-transmittance measurements taken in the mainstem river a few miles below the dam range from 58 to 75 percent absorption. Water in this range is exceptionally clear, and rays of sunshine that induce plant growth can often reach the river bottom. Measurements taken at the Little Colorado River during flood stage indicated a light-transmittance value of less than 0.001 percent absorption.[1] These comparative values represent the difference between a river for which subsurface plant growth is possible and one for which it is not.

Even with an overall increase in amount of light penetrating the river, a pattern of decreasing light transmittance is evident from the dam downstream. Aquatic productivity decreases with distance downstream as well, a gradient recognizable in three distinct river sections: from the dam to the Paria River, from the

Paria to the Little Colorado River, and from the Little Colorado to Lake Mead. Two interrelated reasons account for this pattern. First, the action of the river over the great distance traveled from the dam to Lake Mead works to churn up remaining bottom sediments, thereby reducing light penetration and associated aquatic productivity. Second, tributary streams in the canyon contribute sediment to the river, further reducing light penetration.

Water Temperature

The water temperature of the pre-dam river fluctuated in response to changing seasons, generally following a pattern of summer warmth and winter cold. Winter lows ranged from just above freezing to 40° F. Temperatures gradually warmed to 60° or 70° F during early spring and finally reached 75° to 85° F as annual floods subsided during July and August.

The water that now flows into Grand Canyon is drawn from 200 feet below the surface of Lake Powell, a perpetually cold zone where the warming rays of sunlight never penetrate. Summer and winter, the temperature of water released from the dam is approximately 48° F. With distance downstream from the dam, river temperature in the dam-influenced environment increases by only 1 or 2° F in winter and by fewer than 20° F in summer. Thus the total annual post-dam temperature range is 20° F, compared with a range of as much as 50° F before construction of Glen Canyon Dam.

Usually, the primary productivity of aquatic systems increases with an increase in temperature. But since temperature usually increases at the same time that light availability increases from winter to summer, it is difficult to determine the precise influence of each factor. Most aquatic organisms have a specific set of temperature tolerances, tolerances that are particularly critical for the onset of reproductive activities or life-cycle transitions. Some organisms can tolerate a wider range of temperatures than others, although their all-important reproductive processes and life-cycle transitions can take place only under a relatively narrow range of temperatures.

Thermal constancy day in and day out, season after season, has resulted in conditions for the new Colorado River not unlike those found in a well-balanced aquarium. The unchanging environmental conditions of the post-dam river are favored by only a few species of aquatic organisms, but those few species are found in abundance. They account for biomass production far exceeding that in more diverse and species-rich environments.[2] The down side of stable temperature conditions in the new river is that species that *require* thermal changes at certain life-cycle stages cannot survive. Various physiological processes have different thermal optima, sometimes even within the same organism. For example, stable, cold water temperatures are adequate for the growth of some native fishes in the post-dam river, but the same temperatures have a negative effect on their reproductive cycles.

Stream-bottom Conditions

An important factor determining the quantity and type of bottom-dwelling insects, crustaceans, worms, and other animals is the nature of the stream bottom itself. The absolute lowest productivity usually occurs where the bottom is composed of fine clay and constantly shifting sand substrates, stream-bottom conditions that were prevalent in the pre-dam river. Streamflow characteristics under pre-dam conditions were characterized by annual floods that deposited vast quantities of silt, sand, and larger materials on the river bottom.

Within a few years after the dam was completed, however, the natural shifting silt and sand bottom materials were largely replaced with an armored bed of gravel, cobble, and larger rocks, primarily in the stretch of river upstream of the Little Colorado River and particularly upstream of Lees Ferry. Without the constant input of sediment, the clear water flows simply washed away the finer silt and sand materials, exposing the coarser bed materials.[3] The benefit of this change to the post-dam river is that the most productive stream bottoms consist of cobbles and gravel. Within this coarser stream-bottom substrate, attached algae flourish in greatest abundance and other benthic (bottom-dwelling) organisms living among the algae find shelter. This new, coarser river bottom is fairly stable, providing abundant, relatively permanent locations for the attachment of aquatic plants and other sedentary aquatic life forms.

Chemical Changes Influencing Aquatic Growth

The dam has also served to stabilize the chemical composition and nutrient load of the Colorado. Before the impoundment of the river, chemical composition and concentrations of ions (dissolved salts) were directly correlated with river discharge.[4] During high flows chemical composition of the water was comparatively dilute, a product of chemically "pure" Rocky Mountain snowmelt. It was, however, higher in calcium and bicarbonate than is the case today. During pre-dam low flows the river was characterized by higher ionic concentrations than are found at present. Today the chemical composition of the river stays relatively constant year-round, with Lake Powell serving as a sink, or repository, for those chemical compounds not found in the post-dam river. Though some tributaries, such as the Little Colorado River, may have a substantially different chemical composition from that of the mainstem river, the relatively high flow and turbulent nature of the Colorado ensures a fairly complete mixing and uniformity of dissolved materials.

Nitrate and Phosphate

Nitrate and phosphate are critical limiting factors in plant growth and, as any gardener knows, are quickly absorbed by living plants. The nitrate and phosphate in the post-dam river come mainly from rainfall and erosion of the landscape. Rainwater, in its passage from sky to earth, bonds chemically to the abundant nitrogen in the atmosphere. Although rainfall is sparse in the Grand Canyon region, it does contribute some nitrogen to the river ecosystem. In contrast, most if not all of the phosphorus comes from sediments and the deeper waters of Lake Powell.

One of the most far-reaching changes in the chemical composition of the post-dam river involves the important nutrient phosphorus. Decreasing concentrations of phosphorus in the lower reaches of the river have been documented by studies in Lake Mead. For years Larry Paulson and his students at the University of Nevada at Las Vegas have been working on issues relating to the aquatic ecology of Lake Mead. Prior to Glen Canyon Dam, this reservoir was recognized as one of the most productive fisheries in the Southwest. The Lake Mead food web included a highly productive community of phytoplankton (microscopic drifting plants), large populations of thread-fin shad (a small, phytoplankton-eating fish), and trophy striped bass that reached their great numbers and large sizes (up to fifty pounds and more) by feeding on shad.

Sometime in the late 1970s the striped bass fishery of Lake Mead deteriorated. By the early 1980s, the few fish caught were indicating conditions of severe starvation. The once-numerous populations of shad were almost nonexistent, and even phytoplankton productivity appeared to have declined. Measurements of phytoplankton productivity after 1963 were ten times lower than those taken from 1955 to 1962. The cause and effect of the declining fishery have been traced by Paulson and his students to the lack of available phosphorus in the lake inflow from Grand Canyon.[5]

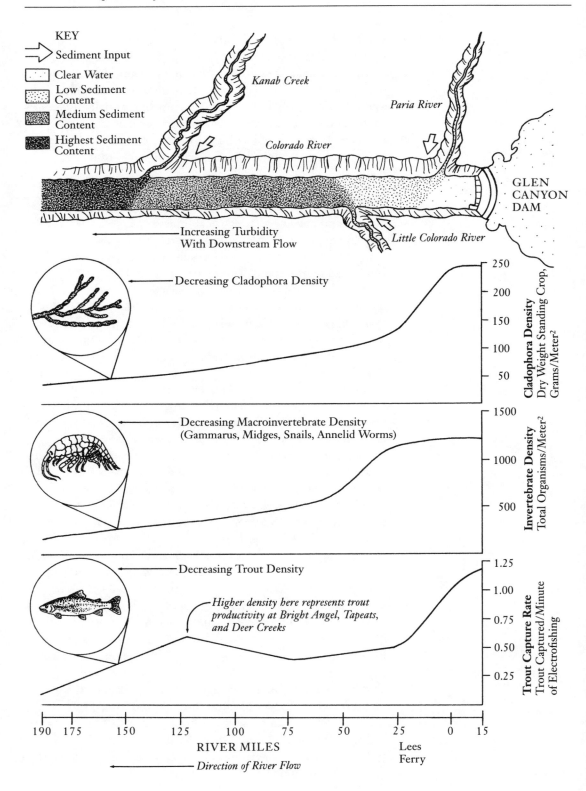

KEY
Sediment Input
Clear Water
Low Sediment Content
Medium Sediment Content
Highest Sediment Content

Kanab Creek

Paria River

Colorado River

GLEN CANYON DAM

Increasing Turbidity With Downstream Flow

Little Colorado River

Decreasing Cladophora Density

Cladophora Density
Dry Weight Standing Crop, Grams/Meter^2

250
200
150
100
50

Decreasing Macroinvertebrate Density
(Gammarus, Midges, Snails, Annelid Worms)

Invertebrate Density
Total Organisms/Meter^2

1500
1000
500

Decreasing Trout Density

Higher density here represents trout productivity at Bright Angel, Tapeats, and Deer Creeks

Trout Capture Rate
Trout Captured/Minute of Electrofishing

1.25
1.00
0.75
0.50
0.25

190 175 150 125 100 75 50 25 0 15
RIVER MILES

Lees Ferry

Direction of River Flow

While searching for the clues to Lake Mead's productivity problems, Paulson turned his investigations toward Lake Powell. He found that the larger upstream lake serves not only as a sediment trap, but as a phosphorus sink as well. Measurements made on the Colorado and San Juan rivers upstream of Lake Powell documented a direct correlation between phosphorus abundance and sediment concentrations.

Under pre-dam conditions, heavy sediment input supplied the phosphorus needed for algal growth when Lake Mead was a productive fishery. Today what little phosphorus does come from Lake Powell is quickly assimilated by abundant algal growth in the clear water below the dam. For Lake Mead, the upstream dam has brought about an almost complete lack of available phosphorus for primary productivity.

Paulson, in cooperation with the National Park Service at Lake Mead National Recreation Area, initiated a program of lake fertilization that has resulted in some improvement in primary productivity over the past few years. Every Labor Day weekend beginning in 1986, Paulson has organized a "fertilizing flotilla" of as many as two hundred motorboats that line up to drag bags of phosphorus across the lake. Preliminary studies indicate that the fertilizer is working: algae are on the increase, the shad seem to be coming back, and Paulson predicts (assuming continued artificial fertilization) that within a few more years trophy striped bass will once again inhabit the waters of Lake Mead.

Although little is known of the nitrogen and phosphorous cycles in the new river regime, these cycles are nevertheless important to aquatic life. Increased light transmittance and, to a lesser extent, a more stable substrate may more than compensate for the lowered phosphate level in the post-dam river and its potential effect on the living systems.

The Aquatic Flora and Fauna

Figure 3.2. The abundance of *Cladophora*, aquatic macroinvertebrates such as *Gammarus*, and rainbow trout decreases with distance downstream from Glen Canyon Dam. The reason for this change is the river's increasing sediment load, which acts to reduce aquatic primary productivity.

Clearly, before Glen Canyon Dam altered the aquatic system, a host of organisms adapted to the high sediment loads and fluctuating temperatures of the "natural" Colorado River. These native organisms were responsible for an unknown, but undoubtedly small, amount of primary productivity in the pre-dam river. There is little question that the primary productivity of the aquatic environment is far greater now than it was then.

Though hundreds of different plant and animal species have thus far been recorded in the post-dam river, only a few contribute substantially to primary and secondary productivity.

The entire story of Colorado River aquatic environments can be told by emphasizing the life histories of these few organisms and how the dam has created conditions perfectly suited to their needs.

Important components of the Grand Canyon aquatic ecosystem include plankton, that mixed group of tiny plants and animals floating, drifting, or weakly swimming through the water column; the attached algae, especially the diatoms; a large, filamentous green alga known as *Cladophora*; a few freshwater invertebrates, including insects such as black fly larvae and chironomid midges; the very abundant introduced crustacean or freshwater amphipod named *Gammarus*; and a small number of other less important plant and animal species.

The Plankton

In ocean, lake, and slow-moving stream systems the plankton community is usually composed of a diverse group of microscopic plant (phytoplankton) and animal (zooplankton) species. Plankton in the new Colorado River is relatively uncommon, and there is strong evidence that most of it found within the mainstream river originated in Lake Powell. Below impoundments it is common to find plankton of lake derivation, with no species of truly riverine origin.[6] Planktonic organisms of all forms simply do not develop as well in flowing water as in the standing and temperature-stratified water of lakes.

Knowledge of the Grand Canyon plankton community before the dam is very limited. In the late 1950s Angus Woodbury, of the University of Utah, was contracted to investigate the aquatic and terrestrial ecosystems that were to be inundated by Glen Canyon Dam. During the two-year study a list of fifty-two algae was compiled, including twenty species of diatoms.[7] A single water sample was taken from the Lees Ferry area in the early 1960s, and the "plankton" identified from it were primarily diatoms.[8]

Many years after the dam had already influenced the aquatic ecosystem in Grand Canyon, the first systematic studies of river plankton were initiated. In an investigation directed by Milton Sommerfeld, of Arizona State University, from 1975 to 1976, fifty-two species (excluding diatoms) of planktonic algae were identified from the mainstem river. Included were two types of yellow-green algae, a golden-brown alga, twenty-three species of green algae, twenty-two species of blue-green algae (now classified as pigmented bacteria, or cyanobacteria), and two species of dinoflagellates. It was found that plankton

CALANOID COPEPODS

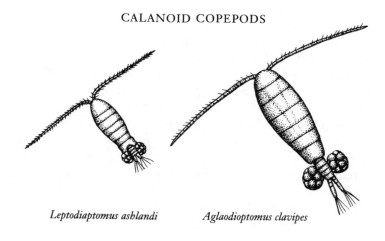

Leptodiaptomus ashlandi   *Aglaodioptomus clavipes*

CLADOCERAN

CYCLOPOID
COPEPODS

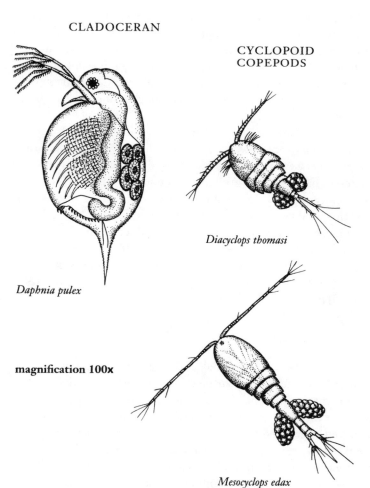

Diacyclops thomasi

Daphnia pulex

magnification 100x

Figure 3.3. Copepods and
cladocerans are microscopic
crustaceans important in the
diets of both native and exotic
fish. Referred to collectively as
zooplankton, these microscopic
animals in the new river come
almost entirely from Lake
Powell.

Mesocyclops edax

composition of the Colorado River is similar to that in Lake Powell and Lake Mead, and the presence of these organisms in the Colorado River in Grand Canyon is due to their being flushed from Lake Powell during discharge through the dam. Sommerfeld also concluded that the phytoplankton densities were relatively low because of the river's swift current, low temperature, and occasional water turbidity.[9]

During the summer of 1980 Loren Haury, of the Scripps Institution of Oceanography, had the opportunity to sample the river throughout its Grand Canyon length for zooplankton. His findings proved that virtually all river plankton was coming from Lake Powell. Moreover, it was all zooplankton, consisting exclusively of microscopic crustaceans representing thirty or more species of calanoid and cyclopoid copepods and cladocerans.[10]

Productivity of zooplankton in Lake Powell and volume and frequency of discharge through the dam play an important role in downstream productivity. Copepods and cladocerans may form a significant dietary component of the native and non-native fish,[11] and they also contribute to the establishment and maintenance of the benthic invertebrate community throughout the entire ecosystem.

Yet the abundance of plankton does not decrease with increasing distance downriver. Unlike the nutrient phosphorus, which decreases in concentration with distance downstream as it is metabolized in plant growth, an abundance of zooplankton apparently remains throughout the entire system, an indication that this food resource is not yet in limited supply. Nevertheless, any management criteria influencing production of lake plankton could immediately and strongly affect the availability of zooplankton as a downstream food resource. The influence of Lake Powell on the downstream river can be dramatic at the most elemental level.

The Attached Algae

In most aquatic systems the algal community is composed primarily of microscopic species. Where bottom conditions permit the attachment of larger species, however, algae can occur in the form of very complex and branched plants. In the armored streambed below Glen Canyon Dam, conditions have been optimum for the establishment of dense growths of *Cladophora*, a green alga. This alga was present in a somewhat limited abundance before impoundment, but it increased dramatically shortly after clear water releases began in 1963.

Worldwide in distribution in both freshwater and marine environments, *Cladophora glomerata*, as it is scientifically known, is perfectly preadapted for life within the dam-regulated waters of the Colorado River. Optimum habitats for the proliferation of this alga include constantly cold, turbulent or wave-swept, clear, and slightly alkaline waters. These conditions all accurately describe water released from Lake Powell. *Cladophora* growth is known to be nutrient-limited, with phosphorus the most common critical nutrient. The extensive stands of *Cladophora* within Glen and Grand canyons undoubtedly account for the depletion of phosphorus in water entering Lake Mead.

Cladophora is commonly and erroneously referred to by fishermen as "moss." This bright emerald, highly branched, filamentous green alga can occur in veritable underwater forests, covering the stream bottom from bank to bank. Each plant is attached securely to stones and cobbles by rootlike structures. Above these attachment structures are filaments with trailing tassels that can reach more than three feet in length. Tassels of algae are easily fragmented by even mild current disturbances, and free-floating filaments are so common in the first 75 miles downstream of the dam that almost every fisherman's lure or fly cast into the swirling waters of the river yields a hook or line full of colorful algae.

The overall primary productivity of the entire river ecosystem is intricately tied to this large and abundant alga, not so much for the substantial productivity of the alga itself, but for the productivity of unicellular diatoms attached to *Cladophora* by the billions. In addition, *Cladophora* beds provide shelter and food for crustaceans and aquatic insects that form other important components of the river's relatively simple food web. A direct and positive correlation is evident between the growth of *Cladophora* and the production of diatoms, aquatic insects, and amphipods. If the large alga is absent, so are the other plants and animals. Thus *Cladophora* is the very foundation of aquatic productivity within the entire riverine ecosystem.[12]

The Diatoms

As a group, diatoms are perhaps the least understood and appreciated living microorganisms occupying freshwater and marine environments. They are frequently included in lists of phytoplankton species encountered in aquatic habitats, because they are found drifting in the water. In the new Colorado River, however, diatoms are most abundant as periphyton, those organisms attached to underwater plant life or other bottom

substrates. Drifting diatoms have, in most cases, come loose
from their preferred substrates.

Like bacteria, diatoms represent some of the most ubiquitous
life forms on earth. Surprisingly, their total net contribution to
the earth's oxygen production is paralleled only by that of trees
and grasses. Diatoms are unicellular microscopic organisms; they
carry out photosynthesis; and they constitute an important
component of the diets of most aquatic invertebrates and many
fish. Studies directed by Dean Blinn and Linn Montgomery, of
Northern Arizona University, have identified more than two
hundred fifty different kinds of diatoms in the Colorado River
and its tributaries.

To the biologist, diatoms are of particular interest because
they are extremely accurate indicators of environmental change.
For example, a few degrees' difference in temperature or
sunlight availability, a slight alteration in water chemistry, or an
imbalance in phosphorus or nitrogen concentrations may
quickly result in a change in the relative abundance of diatoms.[13]
Each species has a narrow range of environmental conditions
within which it will flourish and dominate. With a shift in those
conditions, another species will be favored and eventually
assume dominance. Because of Glen Canyon Dam, the
conditions of the river are relatively stable year-long. Water
temperature, pH (a measure of hydrogen ion concentration),
conductivity, and phosphorus and nitrogen concentrations in
the new river remain constant throughout most of the year. As a
result, only four diatom species constitute the majority of
diatom biomass in the river.[14]

Aquatic Invertebrates

The aquatic insects, crustaceans, and segmented worms present
in the post-dam river were mostly introduced, some long before
the dam was constructed. Many were also planted to provide
food for trout as part of the development of the tailwater fishery
immediately after the dam became operational. Although
relatively few species are present, they are overwhelmingly
abundant. The new river supports large numbers of segmented
worms, chironomid midges, simuliid black flies, and amphipod
crustaceans, the last three being dominant.[15] The oligochaete
worms, earthworms, and some midge species find limited habitat
in shallow riffles where a thin film of sand and organic material
covers the rocks, providing their preferred habitat. Black fly
larvae are found on bare rocks in riffles.

Rhoicosphenia curvata

Achnanthes affinis

Cocconeis pediculus

Diatoma vulgare

**Figure 3.4. Four species of
diatoms are dominant in the
river. They are all representative of aquatic systems
with a relatively high total
dissolved solids content,
including electrolytes that
contribute to higher water
conductivity. The diatoms
pictured here are magnified
more than four hundred
times.**

Gammarus lacustris

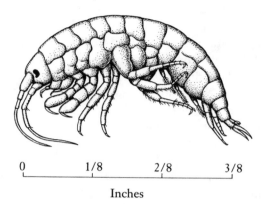

0   1/8   2/8   3/8

Inches

Figure 3.5. Often incorrectly identified as freshwater shrimp, *Gammarus* is an introduced amphipod that is consumed in large quantities by rainbow trout and other fish in the post-dam river.

The tributaries support a far more diverse assemblage of aquatic insects than does the mainstem river. These aquatic insects include mayflies, dragonflies, and damselflies, true hemipteran bugs, diving and predacious beetles, caddisflies, moths, and true dipteran flies. Many of the aquatic insects originating from tributaries are sometimes found in the mainstem river, having been displaced by floods or other disturbances. More than fifty families of insects have been collected from the perennial tributaries and springs draining into the river, where aquatic invertebrate productivity and species richness differs substantially from one tributary or spring to the next.

Living in direct association with the *Cladophora*, and feeding voraciously on the attached diatoms, are the amphipods, sometimes called *scuds*.[16] Although these laterally compressed, side-swimming crustaceans look like, and are distantly related to, shrimp and crayfish, they are not freshwater shrimp, as they are often called. The amphipod dominating the aquatic invertebrate life of the river is known scientifically as *Gammarus lacustris*. Although this tiny animal is partially responsible for the high rate of growth in many of the fishes of the post-dam river, it is not native to the Colorado River in Grand Canyon.

In 1932 fifty thousand amphipods were introduced into Bright Angel Creek as part of an early program to "improve" its fishery. Amphipods were not indigenous to the area, and the attempt to establish them in the creek was one example of uninformed resource management during the early years of the National Park Service. During the introduction amphipods were interspersed throughout the stream with the intention that they would become established as forage for rainbow and brown trout introduced years before. Some amphipods were undoubtedly washed into the Colorado River, but their colonization of the pre-dam muddy waters could not have been successful. Nevertheless, this tiny crustacean was destined to become a pivotal component of the post-dam food web.

Once the dam was in place, the Arizona Game and Fish Department introduced more amphipods, aquatic insects, leeches, and snails into the new clear-water habitats. These introductions were accompanied by annual additions of millions of fingerling trout.[17] The amphipods and other forage invertebrates quickly became well established throughout the river and directly contribute to the phenomenal growth rates and size of trophy rainbow trout.

The Intertidal Zone

Since daily water releases from Glen Canyon Dam are directly related to electrical energy needs throughout the Southwest, river levels fluctuate accordingly. This fluctuation has resulted in the creation of an "intertidal zone" along the shore between the average daily high- and low-water lines. Aquatic and terrestrial productivity in this zone is directly influenced by how much, and on what schedule, water is released from the dam.

The productivity of *Cladophora* and its associated attached diatoms and amphipods is significantly reduced in the intertidal zone. *Cladophora* requires constant inundation to achieve optimum growth. As water fluctuates over the algal beds, sometimes covering and sometimes exposing the vegetation, sharp reductions in growth and survival are effected.[18]

Because of the unpredictable nature of the intertidal zone, no living organisms have become successfully established there. A variety of birds, lizards, toads, and insects have learned to forage for the dead and dying aquatic organisms that are regularly marooned by fluctuating water levels in the intertidal zone. But the ultimate value of the daily tidal drift to terrestrial animals is overshadowed by the tremendous decrease in aquatic productivity caused by the fluctuations in the wetted perimeter of the streambank.

Colorado River Tributaries

The side streams flowing into the Grand Canyon section of the Colorado River are noteworthy in their overall contribution to mainstem aquatic ecology. In addition to their input of aquatic insects and other organic drift elements, many of these streams are refugia, or protected situations resembling pre-dam conditions, for native fish. Pre-dam river conditions no longer exist in the mainstem river, but the perennial tributaries continue to provide a relatively unchanged environment where a variety of native organisms persist in their natural habitats.

The tributaries are primarily of two types. Major drainage systems, like the Paria River, Little Colorado River, and Kanab Creek, are characterized by large watersheds and extensive, deeply entrenched meanders. These three tributaries have relatively low gradients, and they probably contribute most of the post-dam sediment that continues to enter the mainstem river. The other drainage systems entering the river include the relatively short and steep gradient streams, which are usually spring-fed. Though these side canyons can produce spectacular flash floods and occasional debris flows, they are not characterized by regular sediment input.

Approximately fifteen tributary streams exhibit year-long flow into the mainstem river. The cumulative contribution of water from these streams rarely amounts to more than 750 cfs,[19] less than 10 percent of the average post-dam discharge.

Grand Canyon tributaries are, for the most part, as they have been for centuries. Though the watersheds of the major tributaries have been somewhat changed by grazing, mining, agricultural diversions, and some urbanization, these impacts are relatively light when compared to the major changes that have taken place in the mainstem river with the operation of Glen Canyon Dam. We can assume, then, that the aquatic life of the tributary systems has been little changed within the past several hundred years.

The first tributary to enter the river in Grand Canyon is the Paria River, some 15 miles downstream of the dam. The Paria drains a little more than 2,200 square miles of the Escalante Mountains and Paria Plateau of southern Utah and northern Arizona. Except during the driest times of the year, the Paria River contributes sediment-laden water to the mainstem river. During the mid- and late-summer rainstorms of the Colorado Plateau, intense floods down the tributary carry especially heavy loads of suspended materials.

Larger than the Paria and probably the most important tributary to the mainstem river is the Little Colorado River. Entering through a geologically ancient side canyon 61 miles below Lees Ferry, the Little Colorado drains almost 90,000 square miles of eastern and northern Arizona. It is the only major tributary in the Colorado River system whose waters contain a high concentration of sodium salts, the compounds that give the river its ethereal blue color.[20] Like the Paria, the Little Colorado River is also subject to substantial and frequent flash floods. It is the leading contributor of post-dam sediments to the new river. Because of the increased turbidity of the mainstem river below this tributary, downstream aquatic productivity decreases even more dramatically than it does below the Paria.

Kanab Creek, entering the river 143 miles below Lees Ferry, also typically delivers large amounts of sediment during flood flows. Its approximately 3,000-square-mile drainage originates on the Paunsagunt Plateau of southern Utah. Specific studies on decreases in productivity below Kanab Creek have not been performed, although a steady decline in the size of rainbow trout below the tributary (a decline that begins at the Paria River confluence) suggests that riverine productivity as a whole declines as well.

Aquatic Productivity and the Food Chain

The changes in aquatic conditions that have been recorded in the Colorado River in Grand Canyon are of special interest in that the dam has reversed most of the common problems associated with stream deterioration. In the Southwest these problems usually result from erosion and excessive stream sedimentation, caused by unwise land-use practices such as overgrazing, improper timber harvesting, or other human abuses. Repeated documentation shows that the primary productivity of streams decreases strongly with an increase in total sediment volume and turbidity. Such changes are also detrimental to many fish species.

Ironically, Glen Canyon Dam effectively transformed the Colorado River from a muddy stream into a clear one, with a corresponding upward shift in primary productivity. The effect of these changes on organisms dependent on water quality and aquatic primary productivity was as profound as it was sudden. The increased primary productivity provided additional energy that was quickly passed up the food chain to consumer organisms. The fishes of the Colorado River, as the closest consumers on the aquatic food chain, initially responded to these changes in a manner that was completely unexpected by resource managers of the pre–Glen Canyon Dam era.

4 ECOLOGY OF AQUATIC VERTEBRATES

Early in May 1984, 108 miles downstream of Lees Ferry, a group of research biologists from the Glen Canyon Environmental Studies Group was electrofishing. Fishing with electricity is a common research method in which electric current flows into the water between electrodes mounted on a boat. Fish are involuntarily drawn to the anode, or positive electrode, where they can be easily netted.

The group was working just below Bass Rapid when one of them dipped his net into the water and shouted, "A razorback!" When his net surfaced, everyone on the boat strained to see the source of his excitement—he had unexpectedly captured a 2-foot-long razorback sucker.

The reason for the excitement was that this bizarre-looking fish was presumed to be locally extinct in Grand Canyon. It is one of eight original native species thought to have inhabited the canyon before the construction of Glen Canyon Dam. By 1984, the razorback and three other of these fish had apparently disappeared from the river corridor. The last time razorbacks had been seen was in 1977, when members of the fisheries research team from the Museum of Northern Arizona captured three very old individuals in the Paria River near its mainstem confluence.

Relics of the Past

The story of the decline of the razorback sucker reflects the fate of a large proportion of the native fish species that formerly inhabited the river below Glen Canyon Dam: population decline or local extinction. In contrast, exotic fish species have grown in numbers. Thus within historic times the dominant fish of the river through Grand Canyon have completely changed.

The reasons for this shift are all too clear. Principal factors known to have led to the decline of native species in the canyon are introductions of non-native predators and competitors, and revolutionary habitat changes brought about by Hoover and Glen Canyon dams. Glen Canyon Dam's effects on water quality, temperature, turbidity, and discharge have eliminated the river conditions to which the native species were adapted. Furthermore, Lake Mead, positioned at the lower end of the Grand Canyon, provides a reservoir of exotic fish that can easily move upstream into the river.

The prognosis for native fish in Grand Canyon is much the same as it is throughout the Southwest, where native fish populations have been undergoing dramatic changes brought about by human activities since the latter part of the nineteenth century. In addition to water-control projects, aspects of land use and management which have contributed to fish loss throughout parts of the drainage include reduction of riparian vegetation from overgrazing, depletion of groundwater, flow depletions and watershed changes from irrigation and mining, and surprisingly, deliberate attempts to eradicate native fish by poisoning. These changes have subjected the indigenous fish of the Colorado River drainage to tremendous pressures.

The dams and their associated changes to the aquatic ecosystem have also influenced other vertebrate species that spend most of their lives in the river. These animals include beavers, which have dramatically increased in recent times; river otters, which have not been seen for many years; and muskrats, which may be more abundant in the river of the future than they are at present. The lives of these aquatic furbearers doubtless began to change when the first trappers entered the canyon in the early 1800s. In the final analysis, the river's changes have affected both the fish and the furbearers dramatically. Just how dramatically is a subject of much debate, for in this area the historical record is incomplete.

Grand Canyon Fishes: The Naturalized Community

It will never be possible to obtain definitive information regarding pre-dam fish populations of the Colorado River in Grand Canyon. The remoteness of the region and the lack of a recognized need to study the river ensured that no scientific investigations of fish were made before the gates of Glen Canyon Dam were closed. By then it was too late. The limited information that exists on the pre-dam fish community has been discovered in the few historical reports, military journals, and popular accounts from that era. A picture of the species assemblage prior to 1900 has been pieced together from early literature describing the original fish community of portions of the river outside the Grand Canyon.

Before the early 1900s, the dominant fish were probably squawfish, one of three chub species, and flannelmouth and razorback suckers. Channel catfish and carp were introduced into the Colorado River drainage in the late 1800s, resulting in a rapid shift in species dominance. By the time Glen Canyon Dam was completed in 1963, carp and catfish were by far the most common fishes of the river. Many of them invaded from Lake Mead, although both carp and channel catfish were abundant in the Colorado, Green, and San Juan rivers upstream of the site chosen for Glen Canyon Dam.

Channel catfish were far more common in the river before the dam than they are now.[1] From the early 1940s until the early 1970s, it was not unusual for fishermen at Lees Ferry to catch numerous catfish in the range of one to three pounds. Such a harvest was all but impossible by the late 1970s. The channel catfish may have suffered a loss of habitat when the dam's floodgates were closed and the river turned cold, but they remain common in the warmer tributaries.

Just how long catfish have been in the river system is illustrated by the accounts of Ellsworth Kolb. On one of the most exciting river trips of all time, from Green River, Wyoming, to Needles, California, in 1911, Kolb and his brother Emery occasionally used catfish entrails as bait for catching the larger and more desirable Colorado River squawfish.[2] There is little chance that the Kolbs, experienced naturalists and fishing enthusiasts, could have mistaken catfish for any other species. If channel catfish were sufficiently common in Grand Canyon in 1911 to be caught frequently on hook and line, as Kolb suggested, the species must have been introduced many years previously. Native fish species in Grand Canyon may have been coping with this exotic predator since the late 1800s.

Scientific investigations of the fishes of the river did not begin until about 1970, soon after Glen Canyon Dam was built. The first studies were led by Royal Suttkus, of Tulane University. He chartered passage on thirteen commercial river trips through the canyon and eventually collected nineteen species of fish, including fifteen exotics and four natives. Though he was unable to establish the relative abundance of each species, his efforts constitute the first systematic sampling of the mainstem river and tributaries. By that time, the fish community was already a mixture of native and non-native species, with the introduced fishes comprising perhaps 80 to 90 percent of the total biomass. Comprehensive efforts to study the fish populations below the dam began with research sponsored by the Bureau of

Reclamation in 1977, fourteen years after the dam was constructed.[3]

The relative abundance and species distribution of the entire fish community have been changing since these initial studies. Whether or not the fishery will reach an equilibrium is the subject of much speculation. Certain non-native species have become naturalized in their new environs, whereas others were introduced only to disappear in subsequent years. And where suitable habitat exists, some of the original native species continue to thrive.

Indigenous fish species known to have been missing from the system for many years, some certainly declining in numbers even before Glen Canyon Dam, are Colorado squawfish and bonytail and roundtail chubs. The razorback would be listed as extinct in the canyon too, if it were not for the few fish captured since the completion of the dam. This sucker has probably not reproduced in the river since the gates of the dam were closed, and its days in the Grand Canyon appear limited. However, four other native species — speckled dace, bluehead and flannelmouth suckers, and humpback chub — are still present in viable numbers.

If the status quo of the new river is maintained, its future fishery will be a naturalized mixture of native and exotic species. Habitat conditions in the mainstem river continue to favor the non-natives and preclude the successful reintroduction of extirpated native fish. Studies performed over the last fifteen years, however, suggest that the four remaining native fish are precariously maintaining what appear to be stable and reproducing populations. They are sharing the new habitat with at least twenty species of exotic gamefish, minnows, catfish, and carp.

One of the first and most common of the alien species to become established in historic times was the carp. From 1970 through 1978, carp constituted 70 to 80 percent of all fish captured. They were often found congregating near springs and tributaries in schools of five hundred or more adults. But by the early 1980s, the carp had drastically declined in numbers. The Arizona Game and Fish Department, using the same techniques as in previous studies, found that carp at that time accounted for less than 25 percent of all fish captured, and no large schools were encountered.[4] The reason for the decline of carp is unknown but may be related to the ongoing process of habitat change brought about by the dam.

Interestingly, the youngest carp ever captured in the canyon was only two years old, and no evidence (in the form of larvae

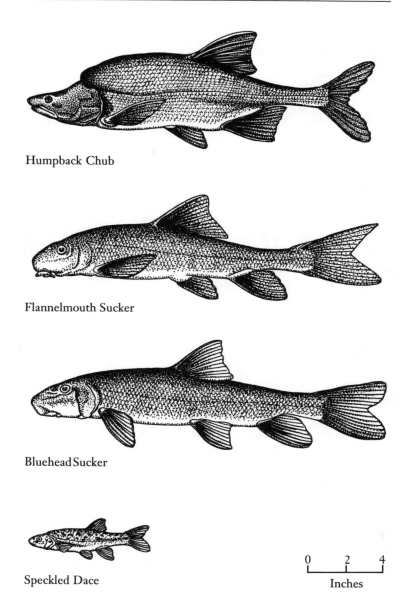

Humpback Chub

Flannelmouth Sucker

Bluehead Sucker

Speckled Dace

0　2　4

Inches

Figure 4.1. The humpback chub, flannelmouth sucker, bluehead sucker, and speckled dace are the only fish species native to the Colorado River in Grand Canyon which persist in stable numbers in the post-dam river.

and subadults) of active reproduction has ever been documented. Carp probably breed in the reservoirs and later disperse into the canyon as adults.

The appearance of non-native fish such as carp and catfish probably helped lead to the original declines in squawfish, razorbacks, and the chubs. Catfish, for example, voraciously prey on larval fish, and carp eat fish eggs and disturb breeding areas.

Not all the non-native species are detrimental to the native ones, however. Many of the exotic species thus far recorded from the new river have been collected or observed so infrequently that they are not important components of the naturalized fishery. Collection records for some species — such as the golden shiner, green sunfish, bluegill sunfish, red shiner, black bullhead, and largemouth bass — in the past two decades probably represent nothing more than isolated captures of nonbreeding escapees from the two major reservoirs, where they are more or less common. These species evidently pass through the turbines of Glen Canyon Dam or disperse upriver from Lake Mead on a limited basis.

Other species for which there are a few records occur as a direct result of deliberate transplants. As late as the mid-1960s little concern was shown for the potential impact of the introductions on the native fishes. The Virgin River spinedace, woundfin, walleye, and coho salmon were all introduced by the Arizona Game and Fish Department shortly after the dam was constructed, in the hope that they would become established.

The woundfin is an endangered species native to the lower Colorado River drainage. It was never found in the Colorado River upstream of the Grand Wash Cliffs, but an unsuccessful attempt was made in 1972 to introduce several hundred individuals into the Paria River. Apparently, the Virgin River spinedace (another native to the lower river) was mixed in with the hatchery woundfins, as it was collected in the Paria River shortly after the transplant.[5] The walleye and the salmon were intended to augment the Lees Ferry fishery, but for reasons not well understood, they failed to reproduce.

The striped bass has been found many times in the new river, but it has not yet become well established. The species has moved upstream into the river since first introduced into Lake Mead in 1969. There is also an abundant supply of striped bass in Lake Powell, and during the spillway bypass discharges of 1983 many were flushed downstream. Fishermen below the dam encountered numerous striped bass shortly after the flood, but subsequent sampling has not produced substantial numbers of this aggressive predator. Since they are dependent on large

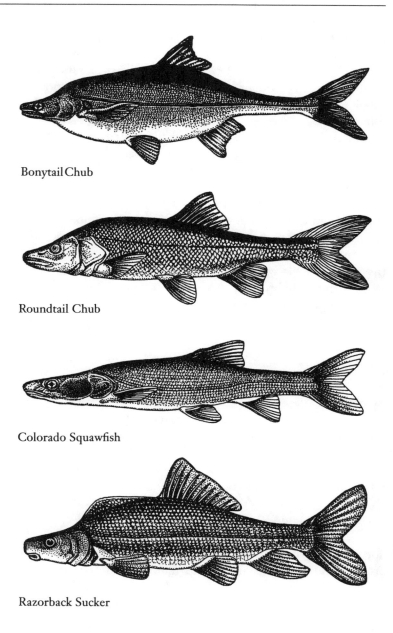

Bonytail Chub

Roundtail Chub

Colorado Squawfish

Razorback Sucker

Figure 4.2. The bonytail chub, roundtail chub, Colorado squawfish and razorback sucker are native fish that are no longer present in that portion of the river through Grand Canyon.

Inches

schools of the thread-fin shad, their establishment in Grand
Canyon is probably precluded by the lack of adequate food.
Striped bass do eat other fish besides thread-fin shad, however,
causing concern that their potential establishment in the canyon
could lead to predation on the endangered species of native
fishes that still persist.

Fathead minnows and Rio Grande killifish are two fingerling-
sized, non-native species that appear well established in some
tributaries throughout the canyon. They are especially common
in the Little Colorado River and Unkar, Royal Arch, and Kanab
creeks. A single record of the Utah chub probably represents an
escapee from Lake Powell, where the species is common.

The remaining non-native species in the river are all trout.
The rainbow trout is now by far the most common fish in the
river, having replaced carp as the dominant species in the late
1970s. Brown and brook trout are found often enough that they
must be reproducing and well established, but the cutthroat
trout seems not to have survived the original stocking efforts in
1978.

Rainbow Trout

By 1920, the National Park Service had implemented an
aggressive sportsfish-introduction program in the Grand
Canyon. By the time the Park Service discontinued its rainbow
and brown trout introductions in 1964, it had planted more than
1.8 million fertile eggs and fingerlings into Bright Angel,
Havasu, Clear, Phantom, Tapeats, Shinumo, and Garden
creeks.[6] No attempts were made before 1964 to stock the
mainstem Colorado River with trout because the relatively warm
and sediment-laden water was not good trout habitat. In the
upper reaches of the new river from the dam to Lees Ferry the
presence of trout can be directly attributed to the Arizona Game
and Fish Department. As soon as the floodgates were closed, the
tailwaters were stocked with thousands of rainbow, brook, and
cutthroat trout. Downstream of Lees Ferry most of the existing
trout population probably originated in the tributaries that were
stocked in the preceding forty years.

From the mid-1960s to the present the rainbow trout fishery
in the Lees Ferry area has attracted fishermen from all over the
country. Dam-induced changes in the river's primary
productivity is perfect for trout growth. Alteration of the basic
ecology of the river, combined with regular and intensive
stocking, has produced a fishery of national renown.

0   5   10
Inches

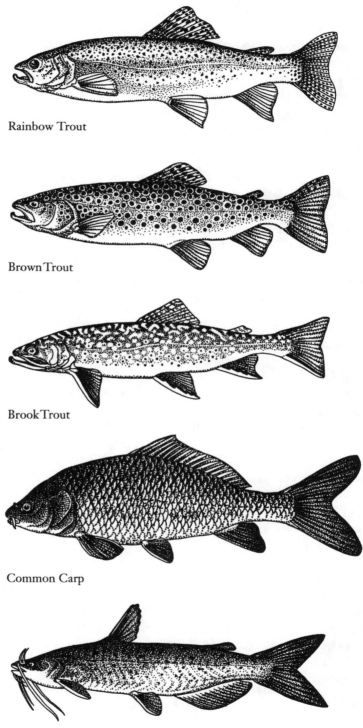

Rainbow Trout

Brown Trout

Brook Trout

Common Carp

Figure 4.3. Twenty exotic fish species, or fish introduced by man, occur in the river. Among the most abundant are rainbow trout, brown trout, brook trout, common carp, and channel catfish.

Channel Catfish

Avid trout fishermen will go to great lengths to fish in a stream where a day's effort results in half a dozen strikes and perhaps one or two fish in the range of one to two pounds. Consider the popularity of a river where a day's fishing yields twenty to thirty strikes, the average fish is two to three pounds, one out of five fish on the line is over five pounds, and every now and again a ten-to-fifteen-pound rainbow trout is landed! Such is the Lees Ferry trout fishery. Mostly, it is a put-and-take fishery where stocking is sufficient to keep up with the harvest. Mainstem river habitats below Glen Canyon Dam are very productive for trout growth, but most attempts by the trout to reproduce are unsuccessful.

Daily fluctuations in river level caused by hydroelectric power production are not conducive to trout reproduction below the dam. Trout build their nests, called *redds*, in gravel bars, usually in shallow areas 1 to 3 feet deep. Typically, a female trout moves onto shallow cobble and gravel bars during high discharges where nests are prepared and eggs are deposited. Several hours later, when power demands decrease and the river level falls, the redds are exposed. Viable eggs die on exposure, and often as not, the female becomes stranded and also succumbs. Occasionally, some reproduction efforts are successful, but they are not sufficient to keep up with the harvest.

The reproduction of rainbow trout is altogether different downriver of Lees Ferry. Fluctuations in flow are attenuated somewhat with distance downstream from the dam, and reproduction in the mainstem river occurs more often. Spawning runs in the tributaries, though, are spectacular. Successful reproduction occurs during most years in Nankoweap, Clear, Bright Angel, Tapeats, and Deer creeks. From November through March these spawning runs often display more than a thousand large trout crowded into the small streams.

Present National Park Service management policies prohibit the introduction of non-native species, and usually every effort is made to correct the mistakes of the past by actively removing alien animals whenever possible. Nonetheless, no programs exist or are contemplated for the removal of non-native fish species from Grand Canyon National Park, primarily because it would be virtually impossible to eradicate the trout without also decimating the remaining native species.[7]

Endemism and Endangered Species

Why are half (four out of eight) of the native fish species of the Grand Canyon threatened or endangered throughout the Colorado River ecosystem, while species brought in from elsewhere proliferate, often at the expense of the native species? A portion of the answer lies in the fact that few native species existed in the ancient Colorado River. This low diversity was due in part to the river's limited geographic area and the general lack of water in environments adjacent to the mainstem and its tributaries. Such geographic isolation results in the development of endemic species, organisms that evolve in a restricted area and are found nowhere else in the world.

The native fish of the Colorado River are mostly endemic, reflecting a long period of relative isolation from other aquatic systems. This situation is not unusual in the desert Southwest, where many native fisheries are characterized by odd-looking relicts, primitive genera, and species with no living close relatives, all indications of millennia of isolation.

The pre-dam Colorado River was not only a relatively limited, closed system, but one marked by enormous variability as well. Wild extremes in river flows, water temperature, and sediment load limited the diversity of fish species that could survive and flourish in the system. Consequently, relatively few fish species evolved in the environment of the entire Colorado River and its tributaries (probably fewer than twenty-five overall).[8] Compare this figure with the almost one hundred species in the Mississippi Basin.

Given this low diversity, it is not surprising that the only native predatory fish of the Colorado River is the squawfish. Fishes in Grand Canyon and elsewhere in the drainage had evolved predator-avoidance mechanisms necessary to survive in the presence of only this single predator. When exotic predators were introduced, the native species, including the squawfish, were unprepared for the consequences. The fate of native Colorado River fishes may be similar to that of many isolated or "island" faunas when subjected to the sudden introduction of non-native predators. Small, isolated populations of animals are frequently decimated by predation and competition with introduced species.[9]

Dam-induced habitat changes in the river may have finished off those species already stressed by introduced predators. Alterations in turbidity, temperature, and water chemistry were the most overt habitat changes, but others such as the decline in backwater areas needed for breeding may have also played a part.

Even with dam-related habitat changes and the introductions of non-native species to the new river in Grand Canyon, three native fish are still represented by what appear to be healthy, reproducing populations. These species are speckled dace and bluehead and flannelmouth suckers. The suckers spawn mostly in the tributaries, where the juveniles remain for two to three years after hatching before venturing into the mainstem river. Breeding habitat for these suckers, especially in Havasu, Kanab, and Shinumo creeks, where spring spawning runs are well known, has not been influenced by the dam.

One of the more interesting of nature's phenomena is the dance of the spawning bluehead suckers. From late April to June the females, somewhat larger than the males, enter the tributaries, where the males await their arrival. As each female scoops out a depression on the sandy bottom, the males are quickly attracted to the movement. With coordinated and rapidly escalating undulations, as many as ten males surround and compress the female between their bodies. This dance can continue for several minutes until the female has released all of her eggs and the water turns white with milt released by the males.

Speckled dace, the smallest of the native fish, can easily be found in all the perennial tributaries. The dace are tiny minnows, rarely exceeding 4 to 5 inches in length, darting in and out of rocks and algae covering the creek bottoms. The species reaches its highest densities where rocky substrates dominate in clear-water, high-gradient streams, such as Shinumo and Bright Angel creeks.

The native dace is not one of the endemic species of the drainage, as it is also widely distributed throughout the western United States. This species is relatively tolerant of introduced predators and the habitat changes brought with stream modification. The dace is the one native species that has increased in density and distribution since Glen Canyon Dam.

Before the dam, dace were probably not very common in the turbid water and shifting sand substrates of the mainstem river. Once the dam resulted in clear water flows, and shallow, rocky substrates appeared along the river margins, dace became more common in the mainstem river. Although dace are now found throughout the river, they cannot reproduce in the cold mainstream water. As with the other native species, the dace require water temperatures in the range of 60 to 70°F for reproduction. Adults forage and grow in the mainstem, but egg laying and larval development take place only in the tributaries.

The squawfish and the bonytail and humpback chubs, endemic to the Colorado River basin, are now federally listed as endangered under the Endangered Species Act of 1973. The razorback sucker may eventually be so designated, and for this reason, most management agencies and fisheries biologists consider the sucker on a parity with the other endangered species. The roundtail chub, although apparently long since gone from the river in Grand Canyon, still persists in abundance along other portions of the Colorado River system.

The Humpback, Bonytail, and Roundtail Chubs

Humpback chub were formerly thought to occur throughout much of the Colorado River drainage in Utah and Arizona. In the post–Glen Canyon Dam era, though, the largest, healthiest population of this species is found within the Grand Canyon at the confluence of the Little Colorado River and the mainstem river. The relict habitat of the Little Colorado apparently provides the ideal conditions for the survival of a small, remnant population.

The earliest record of live humpbacks comes from the river log of some of the first Grand Canyon naturalists, the Kolb brothers. Sometime during May 1911 Emery and Ellsworth Kolb were camped at the mouth of the Little Colorado River and were attracted to the river's edge by a peculiar noise. Ellsworth tells the story:

> Then Emery discovered what it was. On the opposite side of the pool the fins and tails of numerous fish could be seen above the water. The striking of their tails had caused the noise we had heard. The "bony tail" were spawning. We had hooks and lines in our packs, and caught all we cared to use that evening. They are otherwise known as Gila Elegans, or Gila Trout, but "bony tail" describes them very well. The Colorado is full of them; so are the many other muddy streams of the Southwest. They seldom exceed 16 inches in length, and are silvery white in color. With a small flat head somewhat like a pike, the body swells behind it to a large hump.[10]

If it were not for Ellsworth's last comment relating to the large hump, we might be tempted to believe that the Kolbs were indeed catching and eating bonytail chub. The description, however, is that of the humpback chub. The Little Colorado River location, too, is further evidence that the Kolbs were observing humpbacks. The confluence of the Little Colorado

Figure 4.4. Biologists from the Museum of Northern Arizona sample fish populations at the mouth of the Little Colorado River in the late 1970s. Electroshocking techniques to capture fish are less effective in this tributary because of its high salt content. *U.S. Bureau of Reclamation, Lower Colorado Region*

River and a few miles upstream in this warm-water tributary appear to be a humpback chub stronghold where adults and young have always been found in abundance. The importance of the Little Colorado River confluence to the chubs is underscored by the fact that there is no other area within the entire drainage now recognized specifically as a humpback chub spawning and nursery area.[11] For this reason, in 1978 the National Park Service restricted fishing at the mouth of the Little Colorado River, including a stretch of the mainstem Colorado one-half mile upstream and downstream from the confluence.

The humpback chub was not even recognized as a distinct species until it was discovered by Robert Rush Miller, of the University of Michigan, in 1942. While visiting the South Rim of Grand Canyon, Miller was casually browsing through the natural history collections in the park archives when he saw a very unusual, 12-inch-long, preserved fish. An observant park naturalist, Nat Dodge, had captured the fish on hook and line in Bright Angel Creek in 1932. How much longer the humpback would have remained unknown to science if Dodge had not made the effort to pack the fish out of the canyon and properly curate it will never be known.

Because of the humpback's rarity and relatively recent discovery, little is known about its life history. This chub apparently feeds on invertebrate animals and can easily be caught by fishermen on any kind of bait. The maximum size reached by the humpback, and the other two chub species, is about 18 inches. The young, like those of most native species, seem to prefer quiet backwaters where warmer water temperatures and low flow characteristics are optimal for their needs. Humpbacks require water temperatures in excess of 60°F for their spring spawn. These temperatures are no longer available in the new river, except in perennial tributaries. Humpbacks are also found in small, isolated populations in the Green and Yampa rivers and in the Colorado River in Cataract Canyon, although these populations may never have been very extensive.

Figure 4.5. Emery Kolb holds three strings of "Colorado River salmon," or chubs, at the mouth of the Little Colorado River, circa 1911. *Kolb Brothers, Courtesy Special Collections Library, Northern Arizona University*

Historically, there were three very closely related species of chubs (all of the genus *Gila*) thought to occur throughout the Colorado River. The humpback, however, is the only one that continues to exist in the Grand Canyon. Some fisheries biologists believe that the three chub species represent a gradation in adaptation to the swift and turbulent water of the Colorado River ecosystem.[12] In this gradation the bonytail represents the form adapted to intermediate swift waters, the humpback is adapted to the fastest of currents, and the roundtail the slowest. Other known habitat preferences of these species indicate that humpbacks prefer deepwater areas. Bonytails were previously most common in open-river areas in the large river channels, while the roundtails reached their highest densities in the tributary streams. Little is known about the life history or distribution of the bonytail and roundtail chubs in the canyon.

One of the few verifiable records of bonytails from the Grand Canyon consists of nonfossilized remains from flood deposits in Stanton's Cave which are some four thousand years old. Subsequent records are limited to single specimens taken by anglers in 1942 and 1944 from Phantom and Bright Angel creeks. The species was probably still living within the Grand Canyon region immediately after the closure of Glen Canyon Dam. Specimens were reported in Lake Powell in the late 1960s, and a search just below the dam turned up a few in 1970. Isolated captures of adults are still infrequently reported from Lake Powell, but the only areas where reproduction is known to occur are Cataract and Desolation canyons of the upper basin.[13]

By the first half of the twentieth century, the bonytail was in decline throughout much of the lower Colorado River basin, and it had disappeared from the Salt and upper Gila rivers sometime before the mid-1920s. Bonytails were rare in the Colorado and Gila rivers near Yuma, Arizona, in the early 1940s and gone by 1950.[14]

The bonytail is now very rare. Downstream of Lake Mead a few large, old adults persist in Lakes Mohave and Havasu, but natural reproduction has not been documented in many years. Efforts by the U.S. Fish and Wildlife Service to rear young bonytails in hatcheries have been successful. Yet most attempts to reintroduce these hatchery-reared young into the wild have failed.

There are no records of roundtails from the Grand Canyon, in spite of the fact that both roundtail and bonytail chubs were originally described from collections made by Captain Lorenzo Sitgreaves in the 1850s at Grand Falls on the Little Colorado River. Healthy populations of roundtails are known from both above and below the Grand Canyon, where it is assumed they have occurred in the past. They were present in Lake Powell into the late 1960s and were found for a few years below the dam.

Although historical information is lacking, the possibility exists that the bonytail and roundtail chubs were never abundant in Grand Canyon. If at one time they were common, population declines may have preceded the construction of Glen Canyon Dam. All the native chubs are very long-lived species, and surprisingly, some adults recently found in Lake Mohave were almost fifty years old.[15] If the bonytail and roundtail chubs had been present in any numbers before the dam, some individuals probably would have survived as relicts and would have been encountered by now. Curiously enough, there is no record that either species occurred in Lake Mead, although humpback chubs have been found there.

The Colorado Squawfish

The squawfish, once very common, is presumed extinct in the new Colorado River in Grand Canyon. As with the bonytail chub, four-thousand-year-old squawfish remains were found in Stanton's Cave, and they were probably still common in the river in the early 1900s. Describing the squawfish, also known as Colorado or white salmon, Ellsworth Kolb wrote in 1911: "These salmon were old friends of ours, being found from one end to the other of the Colorado, and on all its tributaries. They sometimes weigh twenty-five or thirty pounds, and are common at twenty pounds; being stockily built fish, with large, flat heads. They are not gamey, but afford a lot of meat with a very satisfying flavour."[16]

Limited numbers of small squawfish persist in the upper basin of the Colorado River, primarily in the Green, Yampa, and San Juan rivers. But by the late 1950s, the species' decline in Grand Canyon was apparent. Between 1962 and 1968 one squawfish was collected at Lees Ferry and three others at Glen Canyon Dam. The last verified record from Grand Canyon was a subadult taken at Havasu Creek in 1972 by a fisherman. A remnant population of adults is known to occur at the upper end of Lake Powell at the base of Cataract Canyon, but it shows no indication of successful reproduction.

The squawfish is the largest species of the minnow family native to North America. Although the specimens found in recent years seldom exceed 3 feet and 15 pounds, the largest squawfish on record attained a length of 5 to 6 feet with a weight

Figure 4.6. Emery Kolb holds a large Colorado squawfish caught during the Kolb brothers' 1911 trip down the river through Grand Canyon. *Kolb Brothers, Courtesy Special Collections Library, Northern Arizona University*

estimated between 80 and 110 pounds. Except for a few years during the larval and subadult stages, the species is exclusively a predator on other fish. Squawfish have been reared in captivity, and experimental reintroductions have been attempted in several southwestern streams and rivers. None appears to have been successful.

The Razorback Sucker

The razorback sucker, undoubtedly represented by a few very old individuals like the female captured and released in 1984 at Bass Rapids, probably no longer reproduces in the new river through Grand Canyon or its tributaries. The species will likely be gone by the twenty-first century, even though it is exceptionally long-lived for a fish. Some individuals are known to have reached fifty years of age and more. The old razorback captured below Bass Rapids was a picture of health, with the exception that she was blind in both eyes.

Blindness is not unusual in older suckers that have spent their lives in muddy streams and rivers. They are primarily bottom feeders, and except for a few years during the beginnings of their lives, they have little need for visual acuity. Their food is mostly found by olfactory stimulation, so loss of eyesight is of no particular consequence. That was especially true in the pre-dam muddy water: a sucker with good eyesight living in pre-dam darkness on the river bottom would have no advantage over a blind fish.

Razorback suckers are still found in many of the reservoirs throughout the upper and lower basins, but like other native species, their reproductive efforts are mostly futile. A few reproducing populations are occasionally encountered along the Green and Yampa rivers, and there is evidence of reproduction in the Colorado River near Grand Junction, Colorado. Although young can be produced in large numbers in fish hatcheries, most reintroductions have failed. Because this species is still somewhat widespread, there is hope for its long-term survival, but it is unlikely that the razorback sucker will ever again be a member of the Grand Canyon fish community.

Remembering Mistakes of the Past

Compared to the relatively enlightened resource management principles of today, some practices of the recent past seem incredibly primitive. A common fisheries management practice of thirty years ago was to eliminate the native fish species of a river before introducing exotic gamefish. There were no attempts to eradicate native species from the river in Grand Canyon, although concerted efforts to do so were made in its upstream tributaries. By the early 1960s, the effectiveness of game management agencies at purging streams of native fishes had become dangerously sophisticated.

In 1962 one of the most astounding stream "management" programs ever implemented took place. Sponsored by the U.S. Fish and Wildlife Service and implemented by the game and fish managers of the states of Wyoming, Utah, and Colorado, an attempt was made to eradicate by poisoning all native fishes in 500 miles of the Green River and its tributaries. The impact of this action on the downstream fishery in Grand Canyon is unknown, but there was probably a substantial indirect influence. The reservoir of native fishes that regularly migrated downstream into the canyon must have been severely reduced. The influence of the poisoning program on fishes we recognize today as endangered or threatened is assumed to have been overwhelmingly negative.

A description of the Green River program was dramatically recounted by Robert Rush Miller shortly after his unsuccessful attempt to stop the poisoning:

> The Green River project was the most extensive eradication job ever undertaken, and since the poison traveled through three states it may be justifiably labeled as interstate pollution that was financed, albeit largely unwittingly, by American citizens. Between September 4 and 8, 1962, more than 20,000 gallons of an emulsified rotenone preparation were applied to nearly 500 miles of this river by more than 100 men. The cost of the poison alone exceeded $157,000. Funds for the project were authorized by Congress in June, 1961, after full approval by the Bureau of Sport Fisheries and Wildlife, of the U.S. Fish and Wildlife Service, of the program proposed by Wyoming and Utah. It is admitted by the fish managers themselves that if only a minimum of six years of good trout fishing results from this project, their objective will have been met.[17]

Specific targets of the fish-elimination project were the ten native and nine exotic species known to inhabit the Green River at that time. The intention was that once the river was cleansed of "trash fish," introduced rainbow trout would become established in large numbers. A master plan for the poisoning project had several provisions for preventing "reinfestation and recontamination" by native species. Ironically, the Green River of today has only limited trout fishing, mostly in tailwaters below major dams.

Wildlife management philosophies have matured since that time. Agencies that once fostered the ill-advised poisoning of the Green River (and later the San Juan River in New Mexico) are now making every attempt to "recover" the endangered native species.

One fact is certain — there is no turning back. Changes in the Grand Canyon aquatic ecosystem are so complete that restoration of the native fishery is impossible. What is left is a mix of native and naturalized non-native species which seems to be adjusting into a new equilibrium of its own. National Park Service, U.S. Fish and Wildlife Service, and Arizona Game and Fish biologists and administrators all agree that the only realistic management plan is to attempt to maintain the status quo. Future attempts to manage the aquatic ecosystem of the new river below Glen Canyon Dam will almost certainly give highest priority to protecting the remaining native fishes. More careful attention no doubt will also be given to the exotic fishes that dominate the system, as well as to the aquatic, furbearing mammals that share the river with them.

Mountain Men and Trappers

In the early 1800s a great demand for furs in Europe and the eastern United States was responsible for the legendary influx of trappers, or mountain men, to the river valleys of the West. The trappers primarily came in search of beaver pelts, which were especially prized in the making of fashionable hats for gentlemen. Beavers, and often otters, were almost completely trapped out of existence in some rivers within a few years of the mountain men's arrival.

The Colorado River in Grand Canyon was spared the heavy trapping experienced by most other western rivers. The canyon was mostly unexplored during the early years of the trapping rush, accurate maps were nonexistent, access to the river on horseback was impossible for most of its length, and navigating the river was dangerous and logistically difficult. Notwithstanding the hardships involved, the river was visited by at least one group of trappers in the early 1800s.

In mid-March 1826 a small trapping party on horseback made its way northward along the Colorado River just below the present location of Hoover Dam. Among this group of mountain men was James Ohio Pattie, then twenty-two years old. As he later recounted in a book describing his travels in the Southwest, his party was attacked by Indians and three of his fellow trappers were killed. He and the remaining trappers fled upstream, probably arriving by late March in the vicinity of the Grand Wash Cliffs, the western end of the Grand Canyon. Their exact route past this point is the subject of controversy and may never be known.[18]

Some historians believe that the mountain men continued up the river into the lower Grand Canyon until the encroachment of cliffs at the water's edge forced them to leave the river for the Shivwits Plateau to the north. Once on the Shivwits, they would have paralleled the river by continuing eastward over the Kanab and Kaibab plateaus. Other historians maintain that the trappers traveled along the south rim. Regardless of their exact route, they reached the river's banks again on April 10, possibly near present-day Lees Ferry. The value of this early account is that it does substantiate the small numbers of mountain men who reached the river in the vicinity of Grand Canyon in the early 1800s.

Mountain men remained active in Arizona until the mid-1830s, when trapping all but ceased following a decline in the international price of beaver pelts. In retrospect, any trapping done along the river in Grand Canyon would have had no lasting impact on furbearers, as the activities of the mountain men must have been sporadic and short-lived at most. The next wave of explorers, some of whom were trappers, arrived in the late 1800s and early 1900s to find the river in essentially a pristine condition.

The celebrated voyages of Major John Wesley Powell down the Colorado River in 1869 and 1871–72 left no records of the abundance of furbearers. There is little doubt that Powell and his men were distracted by the danger and uncertainty of navigating an unknown river, and in general, Powell failed to record encounters and items of biological interest.[19] Later explorers, some of whom were experienced trappers, left records that give small insights into the natural abundance of furbearers along the river.

George Flavell was known as "Clark the Trapper" along the Lower Colorado River, his adopted home. In April 1896 Flavell and a companion, Ramon Montez, left Green River, Wyoming, on a voyage down the Green and Colorado rivers to Mexico. The main purposes of the trip were adventure, hunting, trapping, and prospecting. They boated the Grand Canyon in late October, and Flavell's journal recounts some of what they saw.

Like river explorers before him, Flavell must have been simultaneously fatigued by the labors behind him and apprehensive of the rapids he had yet to run during the Grand Canyon portion of his adventure. Although his journal of the trip makes no note of furbearers along the river, Flavell was an experienced trapper in search of furs, and he probably did some trapping in the canyon. The brief time (thirteen days) he spent traversing the canyon suggests either that he did not find trapping profitable enough to warrant a longer stay or that the adversity of the river journey left him little time for trapping. In either case the omission in Flavell's journal implies that furbearers along the river may have been scarce.[20]

Only a few months after Flavell's trip, another river expedition intent on trapping passed through the Grand Canyon. This trip was led by Nathaniel Galloway, hunter, tracker, and trapper. He traveled down the Green and Colorado rivers in the fall and winter of 1896–97, reaching Needles, California, on February 10, 1897. He sold his substantial catch

Figure 4.7. The Birdseye Expedition of 1923 examines a cache of trapper's equipment (note traps at center of photo) discovered near Cave Springs Rapid. Only a few trappers ventured into the river corridor in the 1800s and early 1900s, and records of their trapping success are almost nonexistent. *L. R. Freeman, Courtesy U.S. Bureau of Reclamation, Upper Colorado Region*

of furs there, some of which had been obtained in Grand Canyon, for six hundred dollars.[21]

Other trappers may have traveled the river by boat or horseback into the twentieth century. Or, like Flavell and Galloway, they may have been adventurers who did some trapping on the side while "living off the land." Evidence of trappers along the river in Marble Canyon was found by the 1923 U.S. Geological Survey boating expedition, headed by Claude H. Birdseye.[22] In the cave at Cave Springs Rapid members of the party found a trapper's outfit, complete with traps, cooking utensils, and other equipment. The equipment appeared to have been abandoned, and they could only guess at the fate of the trappers.[23]

The trapping of furbearers along the river apparently continued, at least sporadically, into the 1960s.[24] The overall impact of these activities on native animal populations, however, will never be known. Practically unknown as well is the historic abundance and distribution of furbearers along the river. The completion of Glen Canyon Dam and the changes it caused to the downstream aquatic and riparian ecosystem ensured that the pre-dam status of the three aquatic furbearers along the river would remain a mystery.

Beaver

Beaver occur throughout the river from Glen Canyon Dam to the Grand Wash Cliffs, being most common where riparian vegetation is well developed. They are abundant in three stretches: from Glen Canyon Dam to Lees Ferry, River Miles 40 to 72, and River Miles 166 to 220. Beaver are especially active at dawn and dusk but may be seen throughout the day. After dark the distinctive "alarm" slap of a beaver's tail on the surface of the water signals other beaver that a threat is nearby. Today this threat usually consists of rafters who have unknowingly camped near the beaver's bank burrow.

Riverine habitat for beaver along the mainstem river was not always as favorable as it is today. Vernon O. Bailey made the only scientific observation on the pre-dam status of the beaver in the river when he observed in 1935 that, in general, there were "few suitable places for them and little suitable food."[25] Annual spring floods apparently created unstable conditions for beaver in the river, as well as scouring away seedlings and saplings of their preferred foods, cottonwood and willow trees. Sizeable beaver populations have probably always been found along the larger and more heavily vegetated tributaries, especially Bright Angel and Havasu creeks.

The beaver population apparently began to expand after the completion of Glen Canyon Dam in 1963. This increase in numbers is probably attributable to a combination of factors, including an end to pre-dam spring floods and the post-dam development of extensive riparian vegetation. A November 1977 census in lower Marble Canyon revealed four beaver burrows in a 3-mile reach of river. A conservative estimate of at least two beaver per burrow indicates eight beaver in the 3-mile reach, or 2.67 beaver per mile of river. This density is probably representative of beaver populations in the late 1970s in areas where stands of coyote willow provided suitable food.[26]

Unlike their relatives who build dams and lodges on smaller streams, beaver in Grand Canyon build stick lodges against or on a heavily vegetated beach or else burrow into the banks of the river. The burrow entrance, usually below the average high-water line, leads upward to a small excavated room. Suitable for one or more adult beaver, the room provides a dry refuge in which the young are raised.[27] Fluctuating water levels from Glen Canyon Dam occasionally expose burrow entrances when flows are low, revealing telltale beaver sign: scat, tracks, and discarded twigs stripped of their bark.

Remains of limbs or sticks used as food items comprise most of a bank house. Examination of these diggings indicates a variety of preferred foods, including cuttings of willow, tamarisk, honey mesquite, catclaw acacia, cattails, and the tuberous roots of aquatic and riparian plants. Coyote willow is the beaver's staple food in Grand Canyon, although cottonwood trees are preferred in other areas of the Southwest. Beaver in the canyon also use the larger Goodding willow for food, and its scarcity along the river may partly reflect this preference. The last remaining Goodding willows at Buckfarm and Saddle canyons were cut down by beaver in the mid-1980s, and the stand of large Goodding willows near Cardenas Creek is interspersed with the stumps of trees felled by beaver.

The abundance of beaver along the river is ultimately controlled by the presence of their preferred riparian plant foods. And, in turn, the operation of Glen Canyon Dam controls the abundance and distribution of these plants. The flood of June 1983 scoured away many stands of willow and cattail and was apparently responsible for the temporary decline in the beaver population observed in the several years after the flood.[28] The flood also came at the time when young beaver were in the burrows and possibly drowned many young and adults alike. The operation of the dam will continue to be the primary influence on the beaver population for years to come, both through manipulation of their food supply and through the impact or absence of flood disturbance.

River Otters

Otters were so rare along the river by the late 1970s that their very existence was a matter of speculation. The few historical records concerning otters suggest, however, that they may have always been rare in the canyon. Like beaver, otters have probably been present in Grand Canyon throughout the last ten thousand years. This conjecture is substantiated by an otter bone discovered in four-thousand-year-old driftwood deposits in Stanton's Cave.[29]

The earliest historical mention of river otters was by Vernon O. Bailey, of the U.S. Biological Survey. In September 1889 he observed otter tracks on the riverbank, probably near Hance

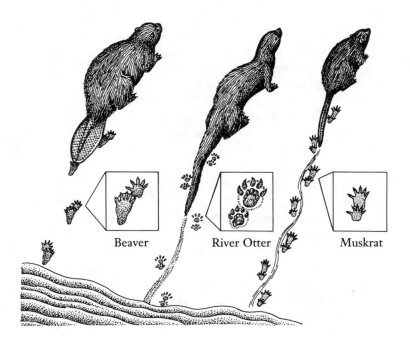

Beaver   River Otter   Muskrat

Figure 4.8. Beaver, muskrat, and river otter are the three species of aquatic mammals found in the river. They look similar when seen swimming at a distance, but on land their appearance and tracks distinctly set them apart.

Rapid. Edward W. Nelson, also of the U.S. Biological Survey, received reports of otter sightings in early 1909 at Lees Ferry. He visited the Lees Ferry site in late summer of 1909, found fresh otter tracks, and saw one otter.[30] Emery Kolb, always observant, saw "occasional otters disporting themselves near our boats" near Diamond Creek in January 1912.[31]

Sporadic sightings of otters from throughout the river corridor continued to be received by the park from 1937 to as recently as 1982. After examining these records, Barry Spicer, of the Arizona Game and Fish Department, concluded that they point to two possible refugia for river otters: the mouth of Bright Angel Creek at Phantom Ranch and upper Marble Canyon near the mouth of Soap Creek. Here four separate sightings of otters or otter tracks were made by reliable observers between 1964 and 1982.

The present-day river in Grand Canyon seemingly represents excellent river otter habitat. Spring floods have been curtailed by the dam, and fish, the otters' favorite food, are abundant. The river habitat also consists of the right proportion of riffle-pool-riffle conditions preferred by otters for foraging.

In the 1980s the Arizona Game and Fish Department began a program of reintroducing river otters into the Salt and Verde rivers of central Arizona. A similar program has been suggested for the Grand Canyon. The National Park Service is investigating this possibility with extreme caution, for fear there might be a relict population of otters still present in the canyon. Should otters still exist, introducing individuals from an alien population could interfere with the genetic makeup of the native group. Nevertheless, there is a strong possibility that someday either native or reintroduced otters could be common in the post-dam river.

Muskrats

The third of the aquatic furbearers found in the river is the muskrat. Essentially an overgrown vole, the muskrat is widely distributed along the larger rivers of Arizona, including the Lower Colorado River below Hoover Dam. Its status in the Grand Canyon section of the river is uncertain. Although the muskrat was traditionally thought to be absent from the canyon, recent sightings may indicate a small resident or transient population.

Edward Nelson saw a muskrat at Lees Ferry in late August 1909, but no other sightings were reported until the 1980s. Tracks seen by Stewart Aitchison at that time were probably those of a muskrat. A dead muskrat was found by biologist and river guide Mike Yard at the mouth of Tapeats Creek on June 9, 1986. The animal had been dead for several days, but the carcass was readily identifiable.

Muskrats, if they are or ever become present on a permanent basis, may also increase in numbers because of the more stable river environment. The growth of riparian vegetation since the construction of the dam should be beneficial to this species. Muskrats eat primarily vegetable matter, including grasses, roots, leaves, and cattails. In turn, muskrats are commonly preyed on by river otters where the two occur together.

Why, in the twenty-five years since the dam has greatly altered the river to the potential benefit of both muskrats and otters, has neither species effectively colonized the new habitat and increased in numbers as beaver have? The answer may well be that both muskrats and otters were present in such low numbers that a sufficient breeding population has never developed. But the full extent of the changes to be brought about by the dam has not yet been manifested, and river otters and muskrats may one day colonize the river of their own accord.

From fish to beaver, the aquatic vertebrates of the new river have been forced to adjust in many ways to the changes brought about by Glen Canyon Dam. The native fish have generally suffered, while native furbearers, particularly the beaver, have fared well. But it took more than just a change in the nature of the aquatic ecosystem to bring about the spectacular rise in the abundance of beaver, an animal that lives in both aquatic and terrestrial worlds. For the beaver, the post-dam changes in the terrestrial, or riparian, habitats on which it depends for food have been as striking as the changes in its aquatic world. A sweeping change in the riparian vegetation along the river was another result of the dam which was almost as sudden as it was unexpected.

The color and temperature of water in the river below Glen Canyon Dam are usually the same as found deep in Lake Powell. *Michael Collier*

Above: The hydraulic jump, or "hole," that formed in Crystal Rapid during the flood of June 1983 was a formidable obstacle to navigation. *Curt Smith*

Right: The large boulders exposed in Hance Rapid at low water are the remnants of debris flows which carried sediment of all sizes far out into the river. *Chris Brown*

Opposite page: The Rio Colorado, or "Red River," ran muddy most of the year prior to the construction of Glen Canyon Dam. The river's sediment load was trapped in Lake Powell after 1963, and the clear water released from the dam exhibited an emerald green color (*top*). The post-dam river only runs red for brief periods when local tributaries bring in sediment (*bottom*), as seen here looking upstream at Soap Creek Rapid. *Robert H. Webb*

Above: Perennial tributaries greatly increase the diversity of the river's aquatic flora and fauna by contributing algae, phytoplankton, diatoms, and aquatic insects not found in the river directly below Glen Canyon Dam. The confluence of the Colorado River (green, entering from lower left) and Little Colorado River (brown, entering from upper left) is shown here. *Collier/Condit*

Left: High, steady releases from Glen Canyon Dam left a "bathtub ring" along the river in the mid-1980s. *Chris Brown*

Right: By controlling annual floods, the dam permitted the development of a new high-water zone dominated by tamarisk. Some native riparian vegetation, including willow and cattail, also became established in the former scour zone. *Bryan Brown*

Below: The riparian zone that existed prior to the completion of Glen Canyon Dam was composed of a thin band of honey mesquite, catclaw acacia, and other native species at and above the river's average high-water line. This old high-water zone vegetation still persists as a relict habitat high above the river. *Steven W. Carothers*

Desert plants such as Utah agave grow just above the river's high-water mark, using lower elevations along the river as a corridor for northward dispersal. *Chris Brown*

Left, top to bottom: The yellow-backed spiny lizard is one of the most common reptiles found along the Colorado River. Large numbers may be seen during summer at the water's edge where they forage on amphipods and aquatic insects washed up onto shore. *Cecil Schwalbe* Beaver are the most commonly seen aquatic mammal in the post-dam river. They feed mainly on the bark of willows and on other native plants, but the introduced tamarisk will also show signs of their use. *Chris Brown* Although bighorn sheep are regularly seen along the river, there is insufficient information to determine whether the dam has influenced their abundance or behavior. *Chris Brown*

Water levels from Glen Canyon Dam may rise and fall over 10 feet daily, occasionally leaving rafts stranded as a result of low water. *Chris Brown*

Havasu Creek exhibits a constant flow throughout most of the year, allowing dense riparian vegetation to develop right down to the water's edge. If Glen Canyon Dam ceased its fluctuating flow regime and adopted a more constant schedule of near-steady water releases, riparian vegetation could likewise flourish down to the river's edge. One potential result of the enhanced riparian vegetation would be the increased abundance of riparian wildlife of all kinds. *Chris Brown*

This portion of the river corridor would have been inundated if the U.S. Bureau of Reclamation had constructed the proposed Marble Canyon Dam. *Chris Brown*

PART III   The Riparian Ecosystem

5

GREEN ON RED
Riparian Plants in a Desert Canyon

Three wooden cataract boats pulled away from Lees Ferry on the afternoon of July 12, 1938, and drifted downstream into Marble Canyon. The expedition, with Norman Nevills as head boatman, had left Green River, Utah, several weeks earlier, passing through Cataract and Glen canyons before reaching Lees Ferry. No expedition like it had ever been launched: its purpose was to conduct a scientific survey of the plants growing in the desert canyons of the Colorado River.

The expedition had been partially commissioned by Elzada Clover, a professor of botany from the University of Michigan. Also on the trip was Clover's young lab assistant, Lois Jotter, enthusiastic about participating in such an unusual project. As the boats left Lees Ferry and approached the entry to Grand Canyon, the level of excitement was high — Clover and Jotter were not only about to become the first women successfully to challenge the rigorous adventure of a full-length Grand Canyon float trip,[1] they were also going to be the first biologists to investigate the river corridor.

Reports of previous river expeditions had made only casual reference to the plants growing along the river. Members of the Ives Expedition of 1857–58 had collected some plants on their upriver journey from Yuma, but only as far upstream as the mouth of Diamond Creek. Major Powell, in the report of his 1869 and 1871–72 expeditions, made only brief reference to vegetation, noting the scattered hackberry trees that marked the high-water line. Robert Stanton also made a few notes on riverside plants during his 1890 trip through the canyon. The systematic observations to be made by Clover and Jotter on this occasion were to be the first and only scientific study of streamside plants performed before the river was influenced by upstream reservoirs. There was not another serious effort for the next thirty-five years, until studies by the Museum of Northern Arizona began in 1972, ten years after Glen Canyon began changing the ecosystem.

The Clover and Jotter Expedition, as it came to be called, got off to a poor start.[2] The intense heat of midsummer and the lack of clean water no doubt helped to aggravate the interpersonal tensions that had been developing among some participants since the trip's beginning at Green River. As a result, Nevills' two boatmen resigned on reaching Lees Ferry and were hastily replaced. Conditions apparently improved somewhat downstream, though Clover still expressed some anxiety over lack of space to store her cacti specimens. Emery Kolb, photographer and boatman, eventually joined the group at Phantom Ranch and participated in the remainder of the trip. They finally reached Lake Mead in late July, having successfully completed the first botanical research expedition through Grand Canyon.

The Discoveries of Clover and Jotter

Clover and Jotter later published their scientific findings in botanical journals.[3] In their articles the botanists stated how the river served as a low-elevation corridor for the upstream dispersal of desert plants. The known ranges of many plant species were greatly increased by these studies, as was the existing knowledge of specific habitats within which riparian plants occurred. Clover and Jotter discovered a previously unknown species of cactus, later named pineapple cactus, and identified several new varieties of other cactus species. Their keen observations did not fail to note the presence of tamarisk, an introduced riparian tree that was already well established along the river.

Figure 5.1. Dr. Elzada Clover (*left*) and Lois Jotter were the first women and the first biologists to travel the entire length of the Colorado River through Grand Canyon, on a 1938 expedition to survey plants. *Bill Belknap, Grand Canyon river guide, courtesy Westwater Books*

By far the most important observation made by the botanists was their conclusion regarding the overall development of vegetation in the riparian zone: "Owing to constantly changing conditions of the talus caused by landslides, and of the river's edge in consequence of periodic floods, there is little climax vegetation [along the river] in the canyon of the Colorado."[4]

Clover and Jotter had hit on the very nature of the existence of riparian vegetation along the river — it was a community subject to constant, predictable, and substantial natural change. Recurring annual flood events scoured vegetation from the lower portions of the riverbanks, with the exception of those rare trees and shrubs clinging to life in protected situations (behind large boulders, in eddies). The native species of plants persisting along the river had evolved under the stressful conditions of flood and scour and were well adapted to survive these disturbances, albeit in small numbers.

The process by which plant communities recover from

disturbance and progress through an orderly series of different types of vegetation toward a mature and stable condition, or climax, is called *succession*. Riparian plant communities, however, occupy regularly disturbed habitats and may never reach a climax state. Some ecologists propose that riparian vegetation under natural conditions represents a "disclimax" community characteristic of disturbed areas and that succession does not occur in this community. The lack of climax riparian vegetation that Clover and Jotter observed along the river was normal for a plant community adapted to natural, predictable flood disturbances.

The natural cycles of flooding that shaped the riparian plant community came to a sudden end with the completion of Glen Canyon Dam. Clover and Jotter might not recognize today's riparian zone. Many of the broad, open expanses of sandy river deposits formerly defining the limits of the annual floods are now covered with dense thickets of tamarisk, willow, arrowweed, and other riparian scrub. A new cycle of disturbance conditions, or lack thereof, has been set into motion by Glen Canyon Dam's control of the river. Under the influence of this modified ecosystem, an entirely new riparian plant community is developing along the river.

Riparian habitats are known to be some of the most productive vegetative associations in existence. They are especially important in the arid Southwest, where rapidly growing, water-loving plants are of limited distribution, usually found only around seeps and springs or lining the banks of the few perennial streams. Originally scarce, southwestern riparian habitats have decreased in extent during the past one hundred fifty years of human development of the region as a result of the depletion of surface water resources.[5] One of the most diligent efforts of wildlife and land management agencies today is to restore and enhance riparian ecosystems. A pleasant surprise was in store for these agencies, then, when they learned that flood control by Glen Canyon Dam had actually *increased* the amount of riverbank vegetation in Grand Canyon. This benefit was not anticipated by either the builders of the dam or the National Park Service ecologists.

The initial vegetation to colonize the old scour zone was tamarisk, an exotic (non-native) species. Classic wildlife management wisdom has contributed to the belief that the tamarisk has no redeeming resource value. In most areas tamarisk is an invading species that normally out-competes native vegetation. But along the river in Grand Canyon, tamarisk invaded a new habitat for which there was no pre-dam analog. In doing so, tamarisk was not competing with a fully developed, existing native plant community.

It is now clear that some vegetation, regardless of its origins, is far better than virtually none. The dam created the opportunity for increased riparian habitat, but ironically, the operation of the same dam now threatens to wipe it away. The future of this new riparian habitat in Grand Canyon is an issue of major concern. Whether the vegetation remains or disappears is a function of management priorities at Glen Canyon Dam.

One River, Three Deserts

The distribution of desert and riparian plants along the river corridor reflects an ecological zone of transition between cold and hot deserts. To the west of the Grand Canyon lies the Mohave Desert; to the south, the relatively lush Sonoran Desert; and to the east and north, the sparsely vegetated desert grassland of the Great Basin Desert. Within the Inner Gorge of the river corridor, just outside the river-influence zone, can be found species of plants considered indicators of all three deserts.

Downstream in the vicinity of Lake Mead the narrow canyon of the river serves as a passageway along which Mohave Desert species penetrate upriver far to the east of their center of distribution. Just upstream of the Grand Wash Cliffs, the western boundary of the Colorado Plateau, Mohave Desert plants intermingle with typical Sonoran Desert species like ocotillo, brittle bush, and barrel cactus. As one would expect, this area also includes abundant representatives of plants common to both deserts, such as creosotebush and white bursage.

The hot desert species gradually drop out, declining in numbers first, then disappearing with distance upstream. Some of the Mohave and Sonoran Desert species persist in abundance as far upriver as the Little Colorado River, but only as isolated occurrences beyond this point. The lower end of Marble Canyon, near the confluence of the Little Colorado River, is considered the eastern limit of the hot desert vegetative association. In Marble Canyon a few Great Basin Desert species such as blackbrush, sagebrush, and rabbitbrush can occasionally be found intermingled with desert grassland. Nowhere within the river corridor, however, are cold desert species sufficiently represented to classify the vegetation as Great Basin desertscrub.[6]

Figure 5.2. Many species of both desert and riparian plants reach the northern or southern limit of their distribution along the river corridor. Selected species illustrate the changing vegetation from Lees Ferry to Lake Mead.

After only one look at the river, Clover and Jotter were struck by the transition: "the river is serving as a southern portal for northward invasion by [low-elevation] plant species . . . such as acacia, mesquite, and barrel cactus [which] extend upcanyon to almost as far as Lees Ferry."[7] They also noticed as they floated downstream that the change from cold to hot desert was not abrupt, but gradual. The progressive loss of grassland and Great Basin species along the river is in sharp contrast to the more rapid addition of Mohave Desert species in lower Marble Canyon and the Muav Gorge. This pattern of downstream change applies to all vegetation of the river corridor, but the gradual loss and replacement of some riparian species near the river is not as dramatic as the wholesale loss and replacement of most desert species on the higher slopes.

Ocotillo, creosotebush, and barrel cactus in the Mohave Desert habitats adjacent to the river below Lava Falls have a relatively lush appearance when compared to the low, sparse

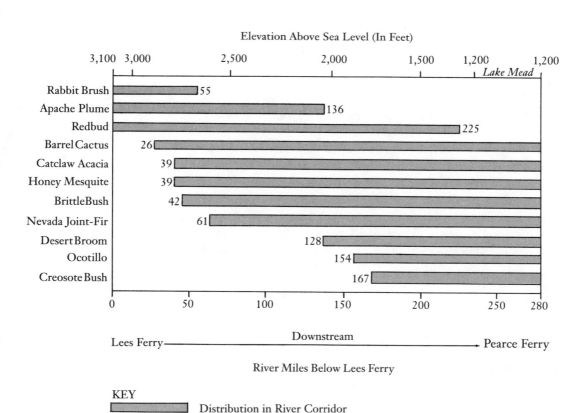

vegetation typical of the Lees Ferry area. The 1,900-foot drop in elevation from Lees Ferry to Lake Mead and the associated rise in temperatures and change in rainfall patterns are apparently responsible for the downstream increase in the amount of living plant material.

Away from the river corridor there exists another type of riparian vegetation, that typical of the seeps and springs. These areas are usually free of recurrent flooding, resulting in a stability that allows each isolated patch of vegetation to develop into a lush oasis. The water availability at the seeps and springs varies tremendously, with the amount of vegetation increasing directly with the amount of water present. Some of the plants typical of these areas are maidenhair fern, crimson monkey-flower, golden columbine, helleborine orchids, and occasionally, sawgrass and beargrass. When conditions allow, trees develop in these areas. Fremont cottonwood, velvet and single-leaf ash, birch-leaf buckthorn, and redbud characterize the more permanent vegetation at larger springs.

The River Corridor Through Time

Alterations to the canyon's riparian vegetation brought about by human interference dominate our thinking now, but events far more powerful produced sweeping changes in prehistoric times. Shifting climatic patterns spanning thousands of years gradually altered the plant community along the river, probably on several occasions.

Scientists have determined, for example, that during the last ice age the climate of the Southwest was cooler and wetter than it was both before that period and since. As a result, woodland communities dominated by pinyon and juniper trees extended to much lower elevations in the canyon. Prehistoric plant remains from Stanton's Cave indicate that about fifteen thousand years ago a woodland of juniper, shadscale, and sagebrush covered the now desertlike talus slopes in Marble Canyon.[8]

The colder, wetter climate caused typical desert vegetation to retreat downstream to lower elevations. The familiar desert plants of the present, such as barrel cactus, brittle bush, ocotillo, and creosotebush, evidently grew only in extreme western Grand Canyon.

The vegetation of the riparian zone also reflected the colder, wetter climate of that time. Riparian plants characteristic of low-elevation, hot deserts probably had to retreat downstream as well. Under those cooler conditions, then, species like mesquite and acacia would not have extended as far upstream as they do today. And riparian plants characteristic of the high-elevation, cold desert would have colonized downstream until reaching the limit of their heat and precipitation tolerances.

The riparian vegetation at the high-water line of fifteen thousand years ago apparently was composed of redbud, scrub oak, netleaf hackberry, and Apache plume — the species we find today almost exclusively limited to the upper reaches of Marble Canyon. Only willows and the occasional cottonwood, almost equally well adapted to conditions during and after the ice age, would have occurred throughout the riparian zone of the river corridor then as they do now, and then only where they were protected from scouring floods.

The climate of the Southwest became warmer and drier after the ice age, and the vegetation of the river corridor has come full circle. Desert and riparian plants favoring a colder climate were forced to contract their range upstream. At the same time, plants of the hot deserts were actively colonizing upstream to the limit of their precipitation tolerances and the appropriate number of frost-free days.

Development of a New Riparian Zone

The desert and riparian vegetation of the river corridor changes not only with increasing distance downstream from Lees Ferry, but with increasing distance upslope away from the water's edge. The water's edge, though, is relative. Seasonal flooding under pre-dam conditions could cause water-level fluctuations of more than 30 feet during the course of a year. The average high-water line of pre-dam floods was generally responsible for the maintenance and distribution of different types of vegetation along the river. Before the dam, three primary belts of vegetation had developed as linear zones running parallel to the river. The appearance and species composition of each linear band were dramatically different, even though each vegetative zone was within a few feet in elevation of the others.

The lowermost band of vegetation, closest to the pre-dam river, was in the scour zone. The scour zone was subject to catastrophic disturbance from the annual floods that swept through the canyon. As a consequence, the only plants able to colonize this zone were either short-lived, ephemeral grasses and herbs or woody species in protected situations more conducive to coexisting with a regime of flood and scour. The grasses and herbs that quickly sprouted in the scour zone after each spring flood were usually washed away by the late summer floods, as were any willows and tamarisk unfortunate enough to

PRE-DAM (BEFORE 1963)

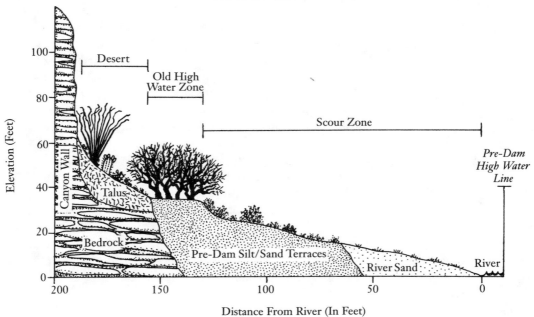

POST-DAM (AFTER 1963)

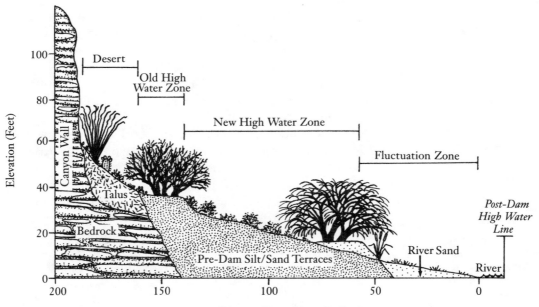

have germinated too close to the water's edge. Battered and stunted specimens of seep-willow, desert broom, tamarisk, coyote willow, and Goodding willow grew sparingly near the upslope limit of the scour zone, nowhere forming dense thickets. However, the presence of well-established willows up to 40 feet in height in some protected areas of the scour zone were mentioned by Clover and Jotter.

The middle band of vegetation along the pre-dam river was composed of riparian plants located just above the scour zone. Referred to as the *old high-water zone*, it was a relatively stable band of riparian shrubs growing at and just above the 100,000-cfs waterline, far enough above the river to escape the scour of annual flooding, yet close enough for nutrient-laden water to lap at its base and soak the roots during an average flood. The old high-water zone of Glen and upper Marble canyons was dominated by redbud, netleaf hackberry, Apache plume, and scrub oak. In lower Marble Canyon these species largely dropped out and were replaced by honey mesquite and catclaw acacia.

The uppermost band of vegetation in the pre-dam river corridor was not riparian in character. Above the river's highest flows and uninfluenced by them, desert vegetation provided a stark background against which the narrow riparian zone stood out like a green oasis. Desert vegetation of the upper river corridor was dominated by pepper grass, Mormon tea, grasses, prickly pear cactus, and a variety of other low shrubs. Adjacent desertscrub in the lower river corridor was composed of typical Mohave and Sonoran Desert species, including creosotebush, ocotillo, Mormon tea, brittle bush, barrel cactus, and cholla cactus.

The construction of Glen Canyon Dam dramatically altered the up- and downslope zonation of vegetation along the river. No changes appeared in the desert zone, which remained out of reach of the river's influence as it always had. Initially, no alterations were apparent in the old high-water zone either. In contrast, the pre-dam scour zone was transformed.

The annual floods that had maintained the scour zone were eliminated by the dam, permitting the unchecked development of a new zone of riparian plants at the post-dam water's edge. Developing at and above the 30,000-cfs level, this new high-water zone was colonized by the same woody plant species that had struggled to survive in its predecessor, the pre-dam scour zone. And this colonization took place rapidly. By 1973, the new high-water zone was a dense riparian thicket 15 to 20 feet high of tamarisk, seep-willow, arrowweed, desert broom, coyote

Figure 5.3. The riparian zone of the pre-dam river was characterized by a broad scour zone inundated by annual floods, above which was a band of honey mesquite, catclaw acacia, and other native plants that comprised the old high-water zone's vegetation (*top*). The riparian zone of the post-dam river is more complex and well developed. The old high-water zone remains unchanged, but the dam allowed the development of a dense new band of vegetation dominated by tamarisk in the new high-water zone where the scour zone had formerly existed (*bottom*).

willow, Goodding willow, and giant reed. Lush cattail marshes sprang up in the backwaters and eddy return channels along a river where marshes had not previously existed.[9]

Aerial photographs of portions of the river corridor have been analyzed to ascertain the rate of change of riparian vegetation in the new high-water zone. A rate of increase of one-half acre per river mile per year was evident from 1965 to 1973. The rate of increase slowed after 1973, but one-quarter acre per river mile per year continued to be added through 1980.[10]

Other types of photography have been used in an equally successful way to highlight subtle, site-specific vegetation changes that have occurred since the dam. The duplication of historical photographs to show environmental changes is a specialty of Ray Turner, a plant ecologist with the U.S. Geological Survey. He realized that a wide variety of historical river photographs had been taken by Jack Hillers of the second Powell expedition, Robert Brewster Stanton, and others over the last one hundred years. Comparison of pre-dam and post-dam photographs taken from the same location at the same scale would clearly indicate trends of vegetation change along the river. Ray and his co-worker, Martin Karpiscak, spent many weeks between 1972 and 1976 painstakingly duplicating historical photographs of the river.[11]

The fruits of their labor were rewarding. Turner and Karpiscak obtained an invaluable sequence of before-and-after photographs documenting vegetation changes in the riparian zone. The detailed nature of some photo pairs was so useful that individual plants could be identified over a period of almost one hundred years, as well as the presence or absence of such features as silt banks, soil cracks, and high-water stains on rocks. This series of photographs indicated an unmistakable trend of increased vegetation in the pre-dam scour zone after construction of the dam, vegetation clearly dominated by tamarisk.

The Colonizers

Tamarisk is an introduced riparian shrub that has become the dominant woody plant in the new high-water zone. Also called saltcedar, it was introduced into the western United States from the Middle East in the late 1800s. Tamarisk spread rapidly and soon became the dominant species of riparian shrub along many rivers of the Southwest. The ability of tamarisk to colonize bare sandbars has resulted in its proliferation throughout the Colorado River drainage. The channels of the Green and upper

Colorado rivers, less confined than the river channel in the
Grand Canyon, have narrowed an average of 27 percent since
the late 1800s, mainly as a result of the spread of tamarisk and its
subsequent stabilizing of sandbars and riverbanks.[12]

Tamarisk probably reached the Colorado River in Grand
Canyon by at least the 1920s. On their 1938 expedition Clover
and Jotter found it throughout the length of the river except for
a portion of Marble Canyon. Tamarisk has a deep taproot, grows
quickly, produces large numbers of seeds, and is well adapted for
life along a large desert river. Like many riparian shrubs,
tamarisk usually develops as single-species stands with little or
no intermixing of other species.

Other introduced riparian plants are found in the river
corridor, though none is as abundant or widespread as tamarisk.
Camelthorn was introduced onto this continent from Asia early
in the twentieth century and had probably begun to colonize the
Colorado River beaches by at least the 1930s. This species
spreads rapidly by means of underground rootstocks.
Camelthorn can quickly take over a bare sandbar, its sharp
spines discouraging any further use of the beach by barefoot
river runners. The increase of this species on some beaches is
becoming a problem.

Russian olive and elm are introduced species that first spread
into the river corridor in the 1970s.[13] These small riparian trees
are inconspicuous and uncommon, as is the introduced Bermuda
grass that is widespread from the dam downstream to Lake
Mead.

Coyote willow is the native counterpart of tamarisk. Like
tamarisk, it is well adapted to the post-dam river and has become
more numerous by rapidly colonizing portions of the new high-
water zone.[14] Coyote willow also tends to grow in single-species
stands with only a few interspersed herbs and grasses on the
ground beneath the willow canopy. The species is more
abundant above Phantom Ranch than below, its downstream
stands not as numerous or continuous as its upstream stands.
This difference is apparently related in part to elevation, which
in turn is responsible for higher temperatures along the lower
river, where beaches dry out sooner and the germination of
willow seeds and the establishment of pieces of willow stems or
rootstocks are inhibited.

The diversity of other native shrubs in the new high-water
zone is high, in spite of the dominance by tamarisk. Arrowweed
is a native riparian shrub that, like coyote willow, is clonal and
shallow-rooted. Unlike the willow, arrowweed is not strictly
confined to the new riparian zone, but was and still is prominent

in the vegetation of the old high-water zone. Arrowweed has invaded large portions of the new high-water zone since the construction of the dam. Because of its clonal nature, individual plants of arrowweed may cover entire beaches.

Goodding willow is the only large riparian tree in the new high-water zone. It has increased somewhat in the post-dam environment but remains uncommon because it is heavily harvested by beaver for food. The only large stand of this willow occurs near the mouth of Cardenas Creek. Goodding willow is long-lived and persistent after it becomes established. There is a gnarled, old individual surviving at Granite Park which has become popular over the years for the shade it provides at a frequently used lunch stop. This tree looks much the same today as it did in 1923 when it was photographed by E. C. LaRue.

Shrub diversity in the new high-water zone is further increased by the presence of four native shrubs in the *Baccharis* group: desert broom, waterweed, seep-willow, and Emory seep-willow. The first two species are similar in appearance and habits, as are the two seep-willows. The final increment to the high diversity of native plants in the new high-water zone has come from the invasion of species that dispersed downslope from the old high-water zone and the desert. Mesquite and acacia occurred in small numbers throughout the former scour zone by the mid-1970s, both as older, reproducing individuals and as young seedlings. Desert plants such as brittle bush and prickly pear cactus were quick to invade the drier habitats within the new high-water zone, playing a small but visible role in the newly evolving plant community.

Overall, the riparian vegetation of the river corridor has gone through a series of four developmental stages from the pre-dam to the post-dam era. The first stage occurred before 1963 when riparian vegetation along the river was limited in extent by natural annual flooding cycles. The second stage began with completion of the dam in 1963 and lasted until approximately 1973. This stage was marked by rapid colonization of the pre-dam scour zone by introduced and native plants and the subsequent development and expansion of the new high-water zone.

Stage three was a time of relative stability for the new riparian zone. Beginning about 1973 and lasting until 1983, this stage was characterized by slowly increasing levels of plant density and diversity due to stable riverflow conditions. Some native plants may have begun to out-compete and replace tamarisk at this time, prompting a few plant ecologists to predict that the new high-water zone would soon reach a post-dam equilibrium of

Figure 5.4. Before-and-after photographs dramatically show the increase in riparian vegetation brought about by Glen Canyon Dam. The upper photo was taken on September 27, 1923, looking upstream from River Mile 204 just above Spring Canyon. The dark band of vegetation bordering the open beaches of the pre-dam scour zone is composed of dense honey mesquite and catclaw acacia. The lower photo, taken of the same location on August 3, 1974, shows the tremendous development of vegetation in the new high-water zone. *Top photo by E.C. La Rue (#633), courtesy U.S. Geological Survey Photo Library, Denver; bottom photo by M. M. Karpiscak, courtesy Museum of Northern Arizona Collectors*

sorts. Glen Canyon Dam was all-powerful, it seemed, and floods
would apparently never return to the Grand Canyon section of
the Colorado River to interfere with the eventual development
of climax vegetation in the riparian zone.

Stage four was ushered onto the scene by the dramatic events
beginning in June 1983, when an uncontrolled flood from the
dam set off a chain reaction that removed all possibility of a
dynamic equilibrium for riparian vegetation in the near future.

Post-Dam Flooding and Vegetation Change

The flood of June 1983 swept through the new high-water zone
like a tidal wave. When the peak of the floodwater finally
receded in late July, the flood's immediate effect on riparian
vegetation was all too clear. Nearly 50 percent of all riparian
plants below the 60,000-cfs water line had been killed. The
cause of this mortality was a combination of factors, including
drowning by continued inundation, burial under newly
deposited sandbars, and the washing away of entire trees and
shrubs. According to various estimates, the new high-water zone
was reduced in extent by 30 to 50 percent.[15]

Floods are the most common and recurring form of
disturbance along rivers. In the final analysis, flood disturbance
also provides the most significant long-term influence over
riparian plant communities. The flood of 1983 temporarily
suspended the expansion of the new high-water zone and, more
importantly, changed the direction of future plant development.
The outcome of this new direction is as yet unknown, but it is
being selectively brought about by latent effects of the flood,
which act to promote some species at the expense of others.

Some riparian species were more susceptible to scouring and
therefore were more easily removed by the floodwater. Species
with deep taproots generally experienced a lower mortality rate.
Tamarisk, Goodding willow, mesquite, and acacia were foremost
among those species able to resist removal by scouring. In
contrast, desert broom, waterweed, and seep-willow are shallow-
rooted species and were unable to survive as well during the
flood. The highest mortality was experienced by the very
shallow-rooted, clonal species such as arrowweed, coyote willow,
giant reed, and cattail. Cattail marshes declined more during the
flood than any type of vegetation in the new high-water zone,
and more than 95 percent of all marshes were lost to scouring.
Of those species that survived, there was differential survival in
different habitats. For example, the mortality of tamarisk was

highest on cobble substrate, moderate on sand, and lowest among boulders or bedrock.[16]

Different species of riparian plants also responded in various ways to long-term inundation by floodwater. Most riparian species are well adapted to survive a brief inundation but are unable to remain submerged for long periods of time. Among all species, nearly 40 percent of those plants remaining below the 60,000-cfs water line had drowned by the time the 1983 flood receded. Tamarisk, giant reed, and coyote willow are very tolerant of inundation, and only a small percentage of their populations drowned.[17] Some tamarisk were known to have been submerged for nearly 850 days through the high-water years 1983 to 1986, yet they survived.

The story of flood-induced mortality in plants of the new high-water zone has been revealed by Larry Stevens, an ecologist from Northern Arizona University. In the late 1970s he had established some study sites along the river. He carefully monitored them each year with the help of his colleague Gwen Waring. Their meticulous work made possible a before-and-after comparison of the effects of the high water.

The length of time over which flooding persisted and the depth of the floodwater were the primary factors that Stevens and Waring found had influenced plant mortality. But differential mortality rates were only one factor that helped to determine the new direction of plant development in the post-flood recovery. The scour and drowning of plants greatly affected species composition and the visual appearance of the new high-water zone, but in the long term the fate of mature plants is not as important as seedling establishment immediately following the high water. And as might be expected, seedling establishment and survival varied widely among riparian species and habitat.

Unlike tamarisk, coyote willow is a clonal species that spreads primarily by means of underground roots. An entire beach several hundred feet long may be covered by a single individual, each above-ground "plant," or stem, a genetically identical clone of the others and linked to them by shallow roots. If above-ground stems are scoured away by flood or burned off by wildfire, the underground root can sprout along its length. In addition, willow stems that are carried downstream, either in high water or as the cuttings left by beaver, may sprout to form a new clone once they are washed ashore. Coyote willow can also reproduce by means of seeds, but this mode of reproduction is rare in the Grand Canyon.

Coyote willow partially abandoned its strategy of single-species stands after 1973, when it aggressively began to colonize adjacent tamarisk stands. The extent and success of this colonization through the early 1980s suggest that coyote willow might be able to out-compete tamarisk along the post-dam river. Some plant ecologists speculate that the willow could eventually replace tamarisk if the stable conditions maintained by Glen Canyon Dam continue.

Floods eliminate mature plants but simultaneously expose patches of bare ground for new germination.[18] The seeds of riparian species such as tamarisk and coyote willow are short-lived, persisting only a few weeks at most during the summer months. The timing of floods is critical for peak seed production to coincide with the perfect combination of water and nutrients supplied by the floodwater.

Tamarisk produces seeds from late April to October, although peak production occurs from mid-May to early June. Coyote willow also produces seeds throughout the growing season, but on a more constant basis. Other native riparian plants generally bear seeds late in the growing season, from July to November. The timing of the 1983 flood, then, favored the establishment of tamarisk seedlings. Their advantage was confirmed by the carpet of seedlings that appeared after the flood of 1983 and also after the smaller floods of 1984–86. Tamarisk seedlings were more than five times as abundant as the seedlings of other species.

Other changes in plant-establishment strategies occurred after the 1983 flood, strategies that signal a major shift in habitat use by tamarisk. Whereas the tamarisk had previously colonized and dominated the sandy beach habitats, it was the coyote willow and arrowweed that quickly reinvaded these flood-scoured areas. These native species were able to occupy the beaches rapidly by expanding their underground network of roots in an efficient form of clonal colonization. Tamarisk and some species of *Baccharis*, restricted to reestablishment by seedling production, were found mostly on cobble bars after the flood.

The net effect of the 1983 and subsequent flooding events on the riparian plants has been a general decline in total extent of riparian habitat. This result is most apparent in the losses of arrowweed, the four species of *Baccharis*, cattails, and giant reed. Where these plants previously flourished, coyote willow and tamarisk have invaded, but by the late 1980s the total acreage of riparian habitat had not returned to pre-flood levels.

The general lack of post-flood recruitment in some of the riparian habitat can be linked to declining nutrient levels in the flood-scoured sandbars. Inundation of beach habitats and the general disruption of these alluvial substrates during the high-water years was damaging to long-term soil-nutrient levels. Sandy soils are relatively poor substrates for plant growth to begin with, as they are usually low in nutrients and water-holding capacity. The sustained flooding acted to leach the few nutrients from the soil, leaving less fertile substrates than before the high-water years. The riparian zone of the future may be severely limited by low soil-nutrient levels and the effects of that deficiency on plant development, germination, and long-term productivity.

The Future of Riparian Vegetation

The riparian vegetation of the new high-water zone no longer has the option of gracefully proceeding toward equilibrium on its own. Too much has changed, too much has already been lost. The "normal" flow regime that occurred from 1963 to 1982, in spite of several distinct drawbacks, at least hinted at stability and eventual equilibrium, and the development of the riparian community was substantial. The alternative to the preceding twenty years of relative stability, the erratic floods of 1983 to 1986, would eventually deplete and degrade the riparian environment. Continued instability may dominate the river corridor for some time, particularly since the probability of additional flooding is higher now that Lake Powell is generally maintained at or near full capacity.

Limited flooding from Glen Canyon Dam could be beneficial. Carefully controlled and planned high flows could be used in a managed scenario to redirect plant development in the new high-water zone. Occasional, relatively small floods could be implemented to promote the germination of native shrub seeds while increasing soil fertility. For example, the high flows of 1983 to 1986 resulted in eventual increases in tamarisk simply because the floods coincided with seed production in this non-native invader. Most native plants produce their seeds many weeks after tamarisk. If the floods could be "scheduled" for August or September, the germination of native riparian shrubs would be favored over tamarisk. The long-term result would be a shift toward a more diverse riparian plant community dominated by native plants.

The most successful program of managed floods would also coincide with the late-summer rainstorms, when unregulated tributaries deliver large amounts of silt and nutrients to the river. Small, dam-induced floods coordinated with the discharge of nutrient-rich floodwaters from the Paria and Little Colorado rivers would act to distribute needed nutrients onto the lower margins of beaches.

Occasional, controlled high flows from the dam may also be necessary to maintain the long-term viability of the old high-water zone. The plants in this community were formerly exposed to seasonal pulses of water and nutrients supplied by the pre-dam annual flooding events. The old high-water zone became something of a relict community and may be heading toward complete senescence in the absence of annual watering.[19] Furthermore, without natural flooding, the germination of seedlings in this vegetation zone has declined significantly. The extent of vegetation in the old high-water zone remained relatively stable from 1963 to 1980. After 1980, however, plant cover declined by 10,000 square feet per mile per year.[20]

Riparian vegetation along the river has not yet reached an equilibrium with the operation of the dam, nor does it appear possible in the near future. Slight changes in how the dam is operated, combined with enlightened management of flood flows, have great potential to direct streamside plant community development to the benefit of the entire ecosystem, a tangible and important goal for the riparian vegetation of the future. As with the aquatic habitats, the primary growth of riparian vegetation dictates the kind and abundance of the animal life that can be sustained in the habitat.

Glen Canyon Dam changed forever the streamside habitats of the Colorado River through Grand Canyon. Though the dam may have violated the National Park Service mandate to maintain natural habitats, the basic elements of the change in riparian habitat were positive. There is more riparian vegetation in the Grand Canyon today because of the dam. This increase was an accident, a secondary benefit of water storage and flood control. Recognizing this windfall and operating the dam to encourage and maximize the riparian growth will also result in substantial increases of native insects and wildlife that rely on the vegetation for food and cover.

6

RIPARIAN INSECTS
Ants, Black Flies, Leafhoppers, and More

Alison Honahni's cry shattered the stillness of the hot August afternoon. From the instant the sting penetrated the skin of her ankle, a burning fire began to move up her leg. The other researchers looked up from their work and watched as she sprinted, still in obvious pain, to the river. Alison had been carefully taking notes on the types of food deposited by harvester ants at the edge of their nest when her luck ran out — she had been stung. The neurotoxic poison of this ant is thought to be among the most powerful venoms found in the animal kingdom.[1] Fortunately, the ants are small, and their venom is injected in minute amounts.

Alison immersed her leg in the river until the cold water numbed the immediate effects of the venom. But for the next two days, inflammation and localized swelling around the sting provided a painful reminder of the hazards of her research project. Considering her work, she was fortunate to receive only a single sting. Multiple stings produce even more excruciating pain and can cause severe neurological disturbances. Those rare persons who are hypersensitive to the venom can go into anaphylactic shock when stung, a condition resulting in very serious consequences, even death, when untreated.

Alison's harvester ant experience is shared by thousands of river runners and hikers in the canyon every year. Yet no deaths or serious injuries resulting from ant stings have been reported from the river corridor. That is surprising, given the large numbers of ants coexisting with river runners in the riparian zone.

Harvester ants, like many other invertebrates of the new riparian habitat, have gone through substantial population fluctuations since construction of the dam. These fluctuations are the direct result of the many changes in both aquatic and terrestrial habitats brought about by the dam. Some riparian insects apparently have responded to these changes by increasing in abundance or by altering their behavior or status. Other hardy species that were not historically present may have even colonized the new high-water zone to become naturalized members of the riverine ecosystem. The best-documented story of adjustment by insects to the changes wrought by the dam is that of the harvester ant. The abundance, behavior, and distribution of this ant have been modified in a way that strongly affects river runners.

Harvester Ants, River Runners, and Glen Canyon Dam

Without a doubt, the harvester ant is the one insect that river visitors take an instant notice of and stay away from. As the name implies, the "harvesting" and storage of mostly vegetable food is the ants' primary occupation. Harvester ants take animal foods as the opportunity arises, but they normally rely on seeds of herbaceous plants to provide sufficient nutrition to form permanent nests.

Pogonomyrmex californicus is the scientific name assigned to the harvester ant found in the river corridor. The name means "California bearded ant," and aptly enough, a hand lens reveals a conspicuous beard of long, curved hairs on the bottom of the head in workers.

These ants are prolific. Each year, usually in late summer, winged males and females engage in nuptial flights. Apparently following some elusive natural cue, probably rainfall, a nearly simultaneous nuptial flight erupts among several distant nests. Copulation takes place in flight, the unfortunate males dying shortly thereafter. On her return to ground, the female immediately begins to dig a shallow burrow. Excavating no more than 4 to 5 inches into the sand, she seals the entrance to the new nest as she digs.

The sheltered female then overwinters in solitude. After a few months she lays a dozen or fewer eggs and fertilizes them with stored sperm. Over the next several months she gradually brings the nest's first brood to maturity. Nourished solely by food reserves stored in the body of the new queen, these first workers are small and timid compared to subsequent offspring. Not until early spring do the workers open the nest and begin to forage. If all goes well, and in the majority of these attempts it does not, the nest becomes established and grows. Usually the maximum nest size is under three hundred individuals, although some nests may support thousands of ants.

In the pre-dam era this cycle was repeated countless times, only to fail — all attempts to establish nests in the scour zone of the uncontrolled river were doomed to certain disaster. The ant queens locating themselves in the more stable habitats in the old high-water zone and adjacent desert had a much better chance at

Figure 6.1. Biologists examine a harvester ant nest beside a grove of tamarisk.
Steven W. Carothers

nest establishment. This pattern changed after the dam. The pre-dam scour zone quickly became some of the best habitat available to ants. The substrate of packed sand was perfect for the digging of nest burrows, and the rapidly developing plant community of the new high-water zone provided an abundant food supply.

Systematic observations of biologists provide a twenty-year history of the harvester ant's invasion of the new high-water zone. The ants quickly responded to the vegetation colonizing the river's edge and just as quickly took advantage of additional food resources provided by river runners and the abundance of black flies in the new intertidal zone. Harvester ants were rapidly increasing in numbers at some heavily used campsites as early as 1976,[2] when ant densities were as high as 2.4 nests per 100 square yards. That was an exceptionally high density for an area known to have been free of any permanent ant colonies before the dam. The growing infestation of ants was primarily attributable to the elimination of annual floods, as well as to the accumulation of food scraps left by the increasing number of river runners. The ants were especially fond of recovering balls of greasy sand where the camp kitchen food waste and wash buckets were dumped.

At that time the policies, techniques, and procedures for keeping beach campsite areas clean were in a fairly primitive state. Tons of human feces, waste food, and cooking grease were contaminating beaches each year. As a result, house flies, blow flies, flesh flies, and ants were increasing in abundance, especially at the popular campsites.[3]

Beginning in 1976, the National Park Service implemented a series of requirements designed to prevent the contamination of beaches with feces and garbage.[4] The measures included dumping kitchen waste buckets into the river rather than on the beach and carrying out all organic garbage. The results were immediate. Exposed garbage and feces were no longer seen in the river corridor, and temporary reductions in most of the insect pests soon followed. In the long term, though, some insect pests have increased in numbers for reasons unknown. Harvester ant numbers declined somewhat with the cleaner beaches, but ants are still abundant in the major camps and immediately adjacent to the river.

Stanley Beus, a professor at Northern Arizona University, has led annual river-based research trips through Grand Canyon since 1982. The purpose of his research is to monitor changes in the physical and biological environment of the riverine habitats by repeating experiments and systematic observations each year. Harvester ant density, distribution, food habits, and relationships with recreational activities have been monitored since the project began.[5]

When the Northern Arizona University studies began in 1982, the average ant densities were about 0.80 nests per 100 square yards. This figure represents a marked reduction from the 2.4 nests measured just six years earlier. Some heavily used beaches still showed higher densities, but the general reduction in ants was clear to experienced river runners.

Then the flood flows of 1983 eliminated harvester ants altogether from the new high-water zone and immediately adjacent campsites. Continued high water through the summer of 1984 prevented reestablishment in these areas; as soon as the water began dropping, however, the ants returned. By the spring of 1985, small nests were reappearing in relatively low densities of about 0.27 nests per 100 square yards. Densities in 1986

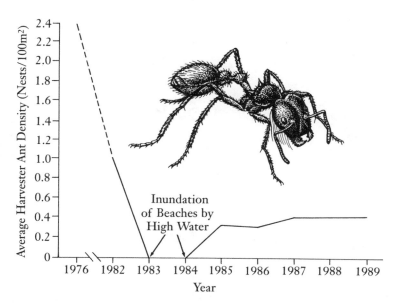

Figure 6.2. Harvester ant densities at camping beaches were very high in the 1970s because of an abundance of food and other organic debris left by river runners. Clean campsite regulations by the National Park Service in the early 1980s led to a decline in the ant population. The flood of 1983 inundated beaches, temporarily eliminating the ants.

increased slightly to about 0.35 nests per 100 square yards, but since then ant abundance has remained more or less the same. The intriguing question is, why have ant numbers not returned to pre-flood levels?

Lack of food is not the reason. Ants may have less human food debris to eat than they had before river runners adopted more fastidious habits, but harvester ants have proven to be highly adaptable. They have discovered yet another feature of the new river environment to exploit: the black fly, which emerges from the post-dam river in numbers that stagger the imagination. And like the new habitat in the old scour zone, the increase in black flies can be directly traced to the influence of Glen Canyon Dam.

The Black Fly Connection

In nature, exclusively carnivorous insects are usually solitary, leading a mostly nomadic existence as they continually search for prey.[6] Colonial insects, on the other hand, are usually vegetarians, because animal foods as a rule are neither sufficiently abundant nor widespread to sustain large colonial populations. As colonial insects, harvester ants generally follow this vegetarian pattern; however, for the ants living near the water's edge in the river corridor, black fly productivity is almost as predictable and stationary a food source as a field of

KEY: FOOD ITEMS

A   Vegetation (seeds and plant parts)

B   Human Food Debris

C   Black Flies

D   Other Insects

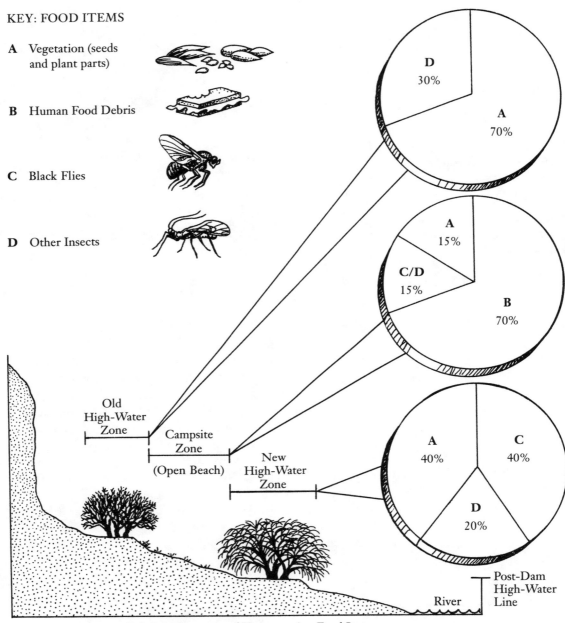

Relative Frequency of Harvester Ant Food Items

Figure 6.3. Harvester ant diets are strikingly different in different habitats along the river.

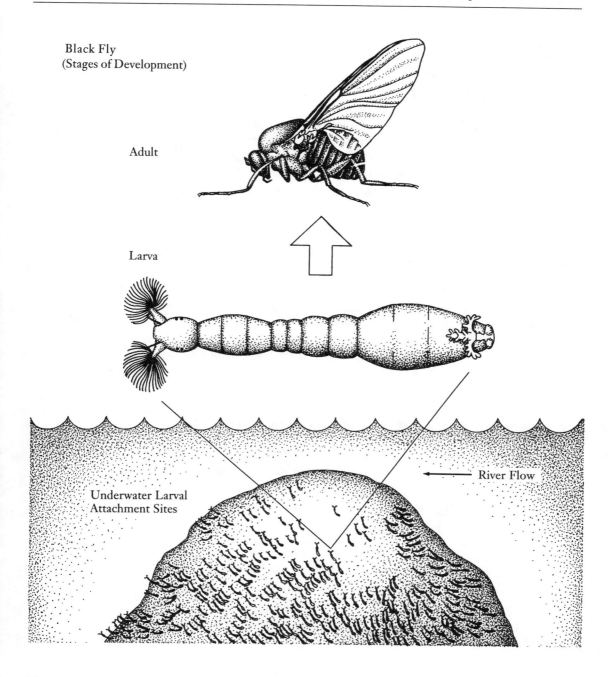

Black Fly
(Stages of Development)

Adult

Larva

River Flow

Underwater Larval
Attachment Sites

Figure 6.4. Black flies are one of the most common flying insects found along the river. The larvae are attached to underwater rocks; adults emerge from the river in tremendous numbers.

grass. Much to the surprise of researchers, a direct link has been established between aquatic productivity, in this case that of the black fly, and use of this new food source by terrestrial ants.

Throughout the spring and summer, when the ants are actively foraging, river-spawned black flies cluster by the millions at the water's edge. Black flies, like the chironomid midges, hatch in the water. But whereas midges simply rise to the water surface and take flight, black flies must first reach land and dry their wings before rising. It is in the intertidal zone of the fluctuating water's edge where black flies concentrate in unbelievable numbers.

The opportunistic nature of the harvester ant's feeding habits has been revealed by monitoring ant nests in different habitats of the river corridor and by examining the types of food items delivered to each nest. Ants are not particular as to living or dead black flies; they take them all, carrying the still-damp flies back to the nest. That occasionally presents a problem, especially during the late summer thunderstorm season when ambient humidity is high. The wet flies apparently cause humidity levels inside the nest to rise above that preferred by the ants. Ant biologists have recorded the massive, if not frantic, removal of hundreds of black fly corpses from a nest. The ants surround the outside of the nest with a ring of dead, wet flies during the full sun of the day. Then, after a few hours of drying, the workers dutifully return the withered bodies to their food stores.

Newly hatched black flies are such easy prey for harvester ants that they sometimes constitute the bulk of the food stored in nests nearest the river's edge. Ants foraging in heavily used campsites opt for a different strategy, carrying mostly human food debris to the nest. Harvester ants in the old high-water zone, when their nests are located far from black fly accumulations and human kitchen wastes, favor a more "normal" vegetable diet of plant parts. Often a significant amount of overlap exists in the ant foraging areas because of the proximity of different habitats at some beaches. For example, in some areas the new high-water zone's vegetation, the campible beach, and the old high-water zone's vegetation are all within 50 feet of each other. This distance is covered on a daily basis by foraging ants from a single nest, and the delivery of food items reflects foraging activities in all three habitats.[7]

These different foraging strategies serve to emphasize the remarkable flexibility of harvester ants' food preferences. Not only have these preferences changed because of the dam, the ants' foraging strategies have been modified by river runners.

Ants seem to be motivated to leave their underground nests and forage in the direction of an active river runners' camp, apparently as a response to the vibrational cue caused by people walking across the beach. The ants' foraging habits and population fluctuations serve well to remind biologists of the sometimes subtle, but complete, change the dam has brought to the aquatic and terrestrial environment of the river corridor.

A Curious Lack of Pest Species

Like harvester ants and black flies, most other arthropods (invertebrates, such as insects and crustaceans, with segmented bodies and jointed limbs) in the riparian zone have increased exponentially since the dam. The dramatic increase in riverside vegetation has provided new and different habitats for insects, while increased aquatic productivity has provided new food sources. Yet even with these increases one universal feature of the new riparian invertebrate community as a whole is that it includes so few truly bothersome pest species. Harvester ants reign as the primary arthropods of discomfort. One leaves the experience of having traveled down the river through Grand Canyon perceiving that, except for the ants, the area is mostly free of obnoxious insects. Nothing can be more agonizing in the great outdoors than the constant attack of biting mosquitoes, no-see-ums, deer flies, and gnats.

Reasons for the general lack of insect pests are not well understood, even though they are cheerfully accepted. Some ecologists suggest that biting pests may increase in numbers as the still-evolving riparian insect community becomes more mature.[8] Evidence indicates that the insect community is still colonizing the new riparian habitats and that additional species of insect pests are certain to become established.

Biting insect pests do appear for short periods in some areas of the river corridor. For example, a small, brown biting beetle, known to certain river runners as the "South Canyon mystery beetle," began appearing in Marble Canyon for a few weeks at a time in midsummer during the 1980s. The beetles seem to occur mostly within 10 river miles upstream and downstream of South Canyon. Although present only in this restricted area, they can dominate a camp on a warm summer night, turning what should have been a delightful experience into a nightmare. Will these beetles become more common with time or was this just a chance irruption? Time and the changing river will provide the answer.

Another pest occasionally swarming into the lives of river runners is the punkie. This tiny fly, resembling a short-legged mosquito, was called "teeth-on-wings" by the early Mormon settlers. Bites by the female punkie are painful and raise an irritating welt that can last for days. Insect repellent is virtually useless as a defense against them. And they never fly alone, but in swarms of hundreds. Punkies can sometimes become so numerous in the Diamond Creek area that seasoned boatmen are driven to tears. If these "teeth-on-wings" should ever become abundant on a regular basis, some river runners and hikers may elect to abandon the river corridor as a popular recreation area.

And why are there few or no mosquitoes in a normal year? Like a few of the other pests, these bloodsuckers occasionally appear but usually do not constitute a problem. The many mosquito species breeding in colder waters of the high country on both rims have yet to colonize what appears to be suitable habitat along the river.

A common denominator in the life cycle of both the biting punkies and mosquitoes is the need for some stability in the river-edge habitat. Mosquito larvae require calm ponds, backwater habitat, or continuously wetted sand in which to develop, and the punkies mature in wet sand. During the high-water years of 1983 through 1986 there seemed to have been more outbreaks of these insect pests than ever before. Interestingly, during those same years daily fluctuations in the river, common during normal operations, were rare. The river ran high and steady, creating backwater areas and maintaining a relatively stable belt of wet sand at the shoreline. High steady flows, or perhaps just constant flows, may create habitat conditions favorable to the proliferation of insect pests.

The Diversity of Riparian Insects

A few other potentially bothersome arthropods live along the river corridor, but they are rare and only occasionally seen. Two species of scorpions, the giant hairy scorpion and the smaller bark scorpion, are formidable in appearance and can deliver a sting that can be more painful than that of the harvester ant. Even more rare than the scorpions are the spiders. Black widow and brown recluse spiders also inhabit the river corridor, safely entrenched in rock cracks, caves, or under driftwood piles, but they are almost never observed.

The relatively few detailed studies on the insect and arthropod inhabitants of the river corridor have all been undertaken by Larry Stevens, of Northern Arizona University. Beginning in 1975, Stevens has been collecting and watching these smallest of colonizers along the river. To date, he has documented the presence of several thousand species, representing 260 families.[9]

Within the group of insects usually referred to as the ground-dwellers are many species of beetles, springtails, and isopterans, or termites. Approximately ten species of termites occupy riparian habitats of the river corridor, but the desert damp-wood termite and *Reticulitermes tibialis* are the most common.[10] Subterranean termites are an abundant and ecologically important insect group found in the river corridor, but they are rarely seen because of their cryptic habits. Subterranean termites feed on dead, buried wood, which they carry to their colonial nests via an extensive network of underground tunnels. Unlike the above-ground ants, which must avoid the heat of the day by suspending their foraging activities until temperatures decline, the termites are constantly active in their cool tunnels.

Termites have microorganisms in their guts that allow them to digest cellulose found in wood. Because dead and down wood decays so slowly in the river corridor and is accumulating in the riparian zones in the absence of floods, termites will play an increasingly important role in the recycling of wood and nutrients into riverside soils. Although little is known of termites along the river, they have colonized portions of the new high-water zone in abundance and may play a critical role in maintaining soil fertility in the potentially nutrient-limited river of the future.

Some of the more obvious insect predators include velvet ants, sphecid wasps, and spider wasps (among them the familiar tarantula hawk). The most commonly seen are the brilliant orange and red velvet ants. They, and their smaller and fluffy-looking white-haired relatives, skitter across beaches in search of their prey of ants and other insects. The tarantula hawk, a deep-purple-bodied large wasp with orange wings, is a relatively common sight in the canyon, but tarantulas are rare. This situation is odd, because the wasp needs the spider — it is within the body of the tarantula that the wasp lays its eggs. Why seemingly far more tarantula hawks than tarantulas inhabit the canyon is yet another invertebrate enigma of the river corridor.

Common among the plant-eating invertebrates are many species of true bugs, each characterized by beaklike, sucking mouthparts. The familiar water boatmen, water striders, and giant water bugs are true bugs, but they are found only in the tributaries and never on the mainstem river. Again, stability of

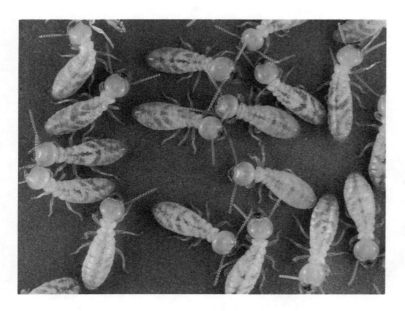

Figure 6.5. *Reticulitermes tibialis* is a subterranean termite that commonly occurs in the moisture and temperature conditions of riparian habitat. This species is abundant in the new high-water zone, where it plays an important role in nutrient cycling and soil formation. *Susan C. Jones, courtesy USDA Forest Service*

the shoreline, as well as the cold water temperatures, may explain why these bugs are largely absent. Brightly colored terrestrial plant bugs, jumping psillids, assassin bugs, delicately patterned lace bugs, leaf-footed bugs, and literally hundreds of other common insect species dwell within the dense vegetation of the river corridor.

Besides ants, other abundant insects of the riparian corridor are hundreds of species of flies, representing the order Diptera. As a group, flies are among the most common of all insects, occupying both terrestrial and aquatic habitats. In the riparian vegetation along the river metallic green and blue blow flies, gray-striped flesh flies, and thick-bodied tachinid flies constantly search for even the smallest amounts of carrion and dung. The ubiquitous house fly is there, too. The house fly reaches its highest densities near humans and their garbage. The species does not exist in the wilderness, except around frequently used campsites.

Tamarisk: Host to Native and Non-Native Insects

When non-native plants such as tamarisk are introduced into one country from another, it is usually in the form of seeds. Sacks of imported grain commonly contain varieties of exotic seeds. Sometimes seeds are inadvertently brought in on clothing, in the fur of animals, or in contaminated soil. Less frequently, mature exotic plants are deliberately brought in by the boatload, as apparently was the case for tamarisk. Surprisingly, the tamarisk did not come from its place of origin alone; insect stowaways came along for the ride. How else could tamarisk-specific leafhoppers and armored scales, also native to the Middle East, have found their way to the Western Hemisphere?

Leafhoppers (especially one named *Opsius stactogalus*) and armored scales are the most common invertebrates found on the plant in the canyon today. These species are so specific to the tamarisk that they rarely disperse to other types of vegetation, and they cannot live for more than a few weeks away from their host.[11] The only way they could exist in this country is by being included in the original introduction of the plant.

Besides the tamarisk stowaways, a few native cicadas and beetles constitute the bulk of the insect community living on tamarisk. The presence of four or five species of invertebrates represents a relatively low diversity. In contrast, the native coyote willow, second in abundance to tamarisk in the new high-water zone, supports scores of insect species. The willow not only has some of its own leafhopper species, but also representatives of several other insect orders and families. Despite this difference in diversity, the actual biomass of insect productivity can be greater on the tamarisk than on the willow. Willows may have a richer, more diverse insect fauna, but tamarisk can produce spectacular outbreaks of leafhoppers.

Why so few species of insects exist in association with tamarisk is not fully understood. Since the plant is not indigenous to the area, the low diversity may simply be a function of the fact that native invertebrates have not had enough time to adjust to its presence and utility. Tamarisk may also contain biochemicals that discourage most plant herbivores.[12]

The insect populations supported by tamarisk appear to be important in the energy flow of the entire riparian community. Massive outbreaks of the tamarisk-specific leafhopper provide a potentially unlimited, if temporary, food source for riparian insectivores. Native carnivorous insects, lizards and toads, birds, and mammals may all benefit from these outbreaks.

Leafhoppers on tamarisk do more than just reach periodic high densities. As phytophagous, or plant-eating, insects, they obtain their subsistence by lancing the plant and sucking its juices. While sucking, *Opsius* secretes a honeydew that rains to the ground, where it attracts other insects. Leafhopper densities on a single grove of tamarisk trees can become so high that their honeydew output is estimated to be in the range of pounds per season![13] The effect of this sugary energy source on the terrestrial food chain could be substantial. The relationship among the tamarisk, its leafhopper, the sugary exudate, and other elements of the riparian community is another new area that awaits further investigation.

The other tamarisk-specific insect of interest is the diaspidid scale. Scale insects are often considered plant pests because, in heavy infestations, they can dangerously stress, and occasionally kill, their plant host. Interestingly, outbreaks of the non-native armored scale on the non-native tamarisk attract a colorful native lady bug known to specialize on scale insects. The predator-prey relationship of the lady bug and the armored scale represents the adaptation of an insect that has been in the Colorado River riparian system for millennia to a new food item in the environment. The lady bug is apparently a fast learner.

The Apache cicada is another native insect that makes use of the tamarisk. Cicadas, like their leafhopper relatives, subsist on plant juices. Adult female cicadas deposit their eggs under the bark of tamarisk and willows, and when the nymphs hatch, they drop to the ground and burrow to the root zone. In the safety of the underground burrow the young cicadas attach themselves to roots and feed for as long as two to three years before emerging. Mature nymphs then dig to the surface and crawl up the first vertical twig or stem encountered. There they immediately split open their chitinous exoskeletons and emerge as winged adult cicadas. Hundreds of thousands emerge within a short time of each other, and the constant buzzing drone of the males can be the dominant summer noise from riparian vegetation.

Along the river corridor Apache cicadas reach very high densities in mid- and late summer, providing many insectivorous riparian birds with prey. The importance of the cicadas to breeding birds may be substantial, especially if the emergence of the insects coincides with the hatching of young birds.[14]

From 1983 through 1985 the deafening chorus of cicadas was not heard in the river corridor. The floodwaters of 1983 apparently drowned the underground nymphs, and it was the summer of 1986 before cicadas were heard in pre-flood numbers.

In the San Pedro River Valley of southeastern Arizona, biologists have noted the presence of approximately sixty-four thousand of these 2-inch-long cicadas per acre of riparian habitat. Surprisingly, more than 85 percent were found in tamarisk.[15] The dominance of the San Pedro River cicadas in tamarisk may simply be related to the fact that suitable native vegetation was not available. Tamarisk also supports high densities of these native insects in the river corridor through

Grand Canyon, although the abundant native willows are preferred by the cicadas for egg-laying.

The fact that the lady bugs and the cicadas are using the tamarisk as both a substrate and food source is an example of the development of a naturalized ecological relationship between native and non-native organisms. Observations relating to the use of tamarisk by indigenous wildlife clearly indicate that the community of the new high-water zone has the potential to evolve toward a healthy and well-balanced ecosystem.

Maturation of the Insect Community

The entire insect community in the river corridor is relatively unstable, if not immature. The influence of Glen Canyon Dam is only now beginning to be recognized, much less understood, by biologists. What is clear, though, is that different types of discharge from the dam, whether flood flows, fluctuating flows, or steady flows, elicit different responses from members of the insect community. To understand fully the consequences of dam operation on the long-term maintenance of the new riparian community, additional research on riparian insects is needed. Most wildlife found in the river corridor is insectivorous in its feeding habits. Changes in water-release patterns which alter insect productivity must, by definition, also influence animals higher up the food chain.

Further signs of the immaturity of the insect community are the increasing arrivals of new colonizers. The bright-orange dragonfly *Libellula saturata*, formerly restricted to the tributaries, has always been fairly rare along the river itself; however, Larry Stevens's observations seem to indicate increasing populations of this and other dragonflies, and he suspects that their full colonization of the new habitat is only a matter of time. The prospect of new insect arrivals to the canyon is of interest, but other wildlife has been attracted, too. Because of the remarkable changes brought about by the dam, other wildlife species have begun occupying the river's new habitats — habitats that were unsuitable for them before the dam became the overwhelming influence on the river corridor.

7

WILDLIFE OF THE RIVER CORRIDOR

Each winter, the bald eagles come to Nankoweap Creek. This annual flight is no coincidence, for the eagles come to feed on rainbow trout during their spectacular spawning runs in the small creek. Bald eagles have not always wintered at Nankoweap, and their recent invasion is another chapter in the story of Glen Canyon Dam and the changing ecosystem along the Colorado River.

Many bald eagles leave their nesting grounds in the Pacific Northwest and Canada every autumn and migrate to the southwestern United States. Several hundred of them linger in Arizona each winter, constantly moving in search of their usual prey of fish. With the onset of spring they migrate back to their northern nesting grounds. A few eagles occasionally visited the river corridor in Grand Canyon before the completion of Glen Canyon Dam, but not in numbers and never for very long.[1]

The preferred food of the bald eagle is fish. Bald eagles eat almost any animal, living or dead, but their foraging specialty consists of diving into flowing or standing water and snatching foot-long fish. When carrion is available, they cluster, vulturelike, around the food until it is gone. The most likely reason they were never common in the canyon was the absence of a readily available food source.

The pre-dam river was inhabited by chubs, squawfish, and bluehead and flannelmouth suckers, but they were largely unavailable to eagles. Sediment-laden water of the pre-dam river undoubtedly made it more difficult for eagles to spot and dive on potential prey. Moreover, there were no large, winter spawning runs in tributaries that would attract and hold numbers of bald eagles. When the native fish did spawn, it was during the spring and summer months, when eagles were already on their northern nesting grounds.

Spawning trout were first observed making large runs in Nankoweap Creek in the mid-1970s. By the early 1980s, large spawns were taking place with as many as fifteen hundred trout in the 15-to-20-inch range crowded into the lower mile of the creek. The spawn at Nankoweap now occurs nearly every year, from November to April, with one or more major peaks of activity from December to March. The trout spawn coincides exactly with the presence of wintering and migratory bald eagles in the Southwest.

The Gift of Eagles

The migration route of many post-nesting eagles appears to be directly over the eastern edge of the Grand Canyon. Once large numbers of trout began using Nankoweap Creek through the winter season, it was only a matter of time before bald eagles discovered the feast and came to stay. Winter raptor surveys of the river corridor were made throughout the late 1970s and early 1980s, but no eagles were recorded. A brief period of delay, or lag time, occurred between the time when the trout spawn first developed and the time when eagles made the discovery and learned to return each winter. Delayed responses to change are not uncommon in wildlife populations, and the developing situation at Nankoweap Creek was no exception.

The first recorded incident of bald eagles at Nankoweap was in the winter of 1985 to 1986, when at least four birds were seen by river runners. Eagles continued to return to Nankoweap each winter thereafter, increasing in numbers every year. By the winter of 1987 to 1988, at least eighteen individuals were present on a more-or-less continuous basis.

Figure 7.1. Bald eagles begin to occur at and near the mouth of Nankoweap Creek in the early 1980s to feed on spawning rainbow trout during winter.

The impressive trout spawn in and at the mouth of Nankoweap Creek has become the focal point for the activities of a small population of wintering and migrating bald eagles. Individual eagles probably leave the creek at times and move up into Marble or Glen Canyon to fish the mainstem river, or soar across the high country of the canyon rims to feed on the carcasses of deer or livestock. But the predictable availability of easy fishing in Nankoweap Creek always seems to drawn them back.

One of the interesting features of this concentration of eagles is the ratio of adults to immatures. The ecological literature on bald eagles indicates that the typical ratio of adults to immatures in a wintering population is two adults to each younger bird, or occasionally even one to one. The ratio at Nankoweap is one adult for every two immatures. This unusually high proportion of young eagles may be a function of the ease with which trout can be captured in the shallow creek. Immature eagles, because of their inexperience, tend to congregate and remain in areas where food is plentiful and easy to obtain. The availability of well-stocked feeding areas, like Nankoweap Creek, enhances the

survival rate of immature birds, as well as adult eagles.

The seemingly unlimited supply of spawning trout in Nankoweap Creek and the relative ease with which they can be captured strongly suggests that the number of bald eagles wintering there will continue to increase. This possibility is reinforced by the historical increase in wintering eagles observed along McDonald Creek in Glacier National Park, Montana. There a spawning run of introduced kokanee salmon which began in 1916 eventually attracted hundreds of migrating eagles and became the largest autumn concentration of bald eagles in the United States outside Alaska.[2]

This recent change in bald eagle distribution in the river corridor is but one of many examples of how Glen Canyon Dam has altered the status and distribution of terrestrial wildlife. A variety of frogs and toads, lizards and snakes, birds, and mammals lives in the riparian zone of the river corridor under the potential influence of the dam. In one way or another the life history or habitat of most of these vertebrate species changed after the dam.

The winter colonization of Nankoweap by bald eagles illustrates the basic pattern of response of most riparian wildlife to the dam — a wider distribution and increase in abundance. This development is a result of the new, highly productive riverside habitat. The degree of change in distribution and numbers depends on how each species makes use of resources and the specific habitat in which it occurs, but the pattern of recent colonization followed by an increase in numbers is widespread.

Wildlife in the New Riparian Habitat

The increase in numbers and species of animals along the river in Grand Canyon is in marked contrast to the overall decline of riparian wildlife in the Southwest over the last several decades. In general, riparian wildlife losses have resulted from a widespread reduction in the quality of free-flowing rivers and associated riparian habitat. Reasons for the loss of riparian habitat are complex but are related to ecological changes in the environment that can be directly linked to urbanization, agriculture, grazing, mining, and the construction of dams and other river-control structures. The construction of Glen Canyon Dam, for example, inundated and destroyed hundreds of acres of lush riparian vegetation along the Colorado River where Lake Powell now occurs. It was quite a surprise, then, that the dam actually allowed an increase in habitat, and subsequent increases in animal populations, downstream.

In nature animals of the same species usually must compete for living space and food. Because of this competition, some animals, usually the young, are forced to leave the immediate area where they were born or hatched, in a process of dispersal. Young animals that disperse into unfavorable habitats may die before adulthood or fall victim to predators, the fate of most animal life. Those animals that disperse into favorable habitats, however, may survive and reproduce to become successful colonizers. Dispersal, then, is the process by which species occupy new areas and expand their ranges over time.

The developing riparian vegetation of the new high-water zone downstream of the dam, with its proximity to water and abundance of food resources, represented favorable habitat and was quickly colonized by animals of all types. Wildlife reached the new high-water zone by dispersing downslope from the old high-water zone or the desert, moving into the river corridor from riparian areas on adjacent tributaries, or in the case of a few determined species, even expanding their ranges hundreds of miles upriver to take advantage of the new habitat.

Birds

The most remarkable long-distance colonizers were the birds. Four species, including Bell's vireo, summer tanager, hooded oriole, and great-tailed grackle, expanded their nesting ranges hundreds of miles upriver as a result of the new habitat created by Glen Canyon Dam.[3] These four species nest only in well-developed riparian vegetation and are relatively easy to find and study in the otherwise arid region along the Colorado River. Biologists have been monitoring the activities of these birds since the early 1970s, and their range expansions have been well documented.

The arrival and upstream colonization of the tiny Bell's vireo are representative of those species that moved into and stayed in the new habitat. Bell's vireo was known to nest only as far upstream as Lava Falls Rapid (River Mile 179) in 1971, eight years after the construction of Glen Canyon Dam. The new high-water zone had become sufficiently well established by that time to allow the vireos to begin colonizing upriver. Vireos like to nest 3 to 4 feet off the ground in the middle of a stand of heavy vegetation, like tamarisk, seep willow, desert broom, and willow — the very species of plants that increased after the dam. Vireos were discovered nesting as far upriver as President Harding Rapid (River Mile 43) in 1982, representing a range expansion of about 135 miles in eleven years.[4] And in the seven years from 1976 to 1982, their numbers more than doubled, from 67 pairs to 135 pairs.

Summer tanager, great-tailed grackle, and hooded oriole reached the upstream limit of their nesting ranges along the Colorado River between Needles, California, and Lake Mead before Glen Canyon Dam was constructed. Once the dam began operation and the new high-water zone had begun to develop, all three species quickly invaded. They had colonized the river corridor as nesting species by the 1970s.

Many other species of birds, although not long-distance colonizers, have moved into the new riparian habitat from adjacent tributaries or the old high-water zone. By taking advantage of the new habitat, they have increased in numbers as well. Most of these birds, such as northern oriole, yellow warbler, and yellow-breasted chat, to name but a few, nest only in riparian vegetation or exhibit a definite preference for it.

Table 7.1 Changes Among Birds Along the River Corridor as a Result of Glen Canyon Dam

Species Increasing in Numbers

Black-chinned hummingbird[a]	Yellow-breasted chat[a]
White-throated swift	Black-headed grosbeak[a]
Willow flycatcher[a]	Blue grosbeak
Ash-throated flycatcher	Lazuli bunting
Violet-green swallow	Indigo bunting
Bewick's wren	Red-winged blackbird[a]
Marsh wren[a]	Brown-headed cowbird
Blue-gray gnatcatcher	Northern oriole[a]
Lucy's warbler	House finch
Yellow warbler[a]	Lesser goldfinch[a]
Common yellowthroat[a]	

Colonizing Species

Extirpated Species

Bell's vireo

Cliff swallow

Summer tanager[a]

Great-tailed grackle[a]

Hooded oriole

[a]This species nests exclusively or primarily in the new high-water zone, indicating that it was either absent from or rare in the pre-dam river corridor.

The black-chinned hummingbird is one of the many summer birds of the river corridor which nests only in riparian vegetation. This tiny hummingbird is one of the least conspicuous species nesting in the new habitat, yet it is the most abundant bird along much of the river. The fact that black-chinned hummingbirds nest only in the dense, lush vegetation of the new high-water zone strongly suggests that they were restricted to tributaries and probably did not even occur as a nesting species along the river under pre-dam conditions.

The new high-water zone also hosts a larger density of nesting birds than does the old high-water zone. A comparison of the numbers of nesting birds per acre in the two riparian zones from 1984 to 1987 showed a significantly greater number of birds in the new habitat, a contrast that was consistent throughout the mid-1980s. From the perspective of maintaining both a high density and diversity of nesting birds in the river corridor, the importance of the new high-water zone cannot be overstated.

The birds that have colonized the new vegetation zone are all native species, with two minor exceptions. European starling and house sparrow are the only two introduced species of birds reported in the river corridor. They are rare and occur only in association with areas of human habitation such as Lees Ferry and Phantom Ranch. A pair of house sparrows even nested in the heavily used river camp across from Deer Creek Falls in the summer of 1975. Both the European starling and the house sparrow were introduced into the eastern United States in the mid- to late 1800s and quickly settled throughout most of North America. Neither species will probably ever become common along the new river.

The willow flycatcher, as the rarest regularly nesting bird of the river corridor, is in many ways the opposite of the black-chinned hummingbird. Willow flycatchers have always occurred in the river corridor, nesting in small numbers in the few willows able to withstand the scouring floods of the pre-dam regime. Reported as common along the river in Glen Canyon in the late 1950s, the willow flycatcher is now either absent or very rare in the few undammed miles of Glen Canyon above Lees Ferry. The species apparently increased in numbers along the river corridor through Grand Canyon because of the recent development of the new high-water zone. The willow flycatcher population in the river corridor fluctuated from seven to eleven breeding pairs during the mid-1980s, making it the largest breeding population of the species persisting in Arizona.

Willow flycatchers declined drastically in numbers throughout the desert Southwest during the mid- to late 1900s and were listed as endangered by the state of Arizona in 1988. Ironically, the largest remaining population in Arizona is in the dam-created new high-water zone of the river corridor through Grand Canyon, where all recently found willow flycatcher nests were in tamarisk, the introduced riparian plant.[5] This situation is only one example of the classic conflict faced by natural resource managers of the National Park Service — what is the proper management solution when a rare or endangered native species comes to rely on non-native vegetation? Should the non-native vegetation be removed and discouraged or enhanced?

The mostly non-native vegetation of the new high-water zone holds a great attraction for nesting birds. Of the fifty-one species of birds known to nest in the river corridor, twelve do so exclusively or primarily in the new high-water zone.[6] Only three species, the phainopepla, Costa's hummingbird, and northern mockingbird, nest primarily in the mesquites and acacias of the old high-water zone, and apparently they have not increased in number as a result of the dam. The twelve species limited to the new high-water zone would be almost entirely eliminated from the river corridor if this new zone were to vanish. The extreme dependence of so many species of birds on the new high-water zone illustrates how its development has increased the overall diversity of nesting birds along the river. This dependence further emphasizes the need to control, if not prevent, flood releases similar to the 1983 event, which greatly reduced the extent of the new high-water zone's vegetation.

Only one species of bird has been lost from the river corridor because of the effects of the dam. Before the dam, cliff swallows raised their young in mud nests clustered in large colonies on rock walls at the river's edge. The change in the sediment balance of the river after the dam, and the resulting lack of mud for nest construction, led to the eventual decline of these birds. Cliff swallows last nested along the river in 1975 and, without mud for their nests, will probably not return.

Violet-green swallows and white-throated swifts, the two most common aerial insectivores along the river, may have greatly increased in density since the dam. These species find virtually unlimited nesting habitat in the cracks and clefts of the canyon walls, and there is strong circumstantial evidence that their primary food has dramatically increased in recent years. Feeding only on the wing, swifts and swallows are exclusively dependent on flying insects. Two dam-related changes have greatly increased the flying insect density directly over the river corridor: the new riparian vegetation and the increased primary productivity of the river itself.

Mammals

Like the swifts and swallows, bats of many varieties may have increased since the dam. Bats, seen by the thousands during summer, are nocturnal predators on flying insects. In the twilight of dawn and dusk numbers of big brown bats, Townsend's big-eared bats, many members of the *Myotis* complex, pipistrelles, and occasional hoary and silver-haired bats can be identified swooping over the river or through the riparian vegetation. All night long the bats are actively foraging on the same insects that support insectivorous birds during the day. The new habitats seem to produce an almost unlimited supply of flies, gnats, and midges for all species of aerial predators.

Terrestrial mammals also responded to the influence of the dam with vigor. Some quickly dispersed into the developing new high-water zone, greatly increasing in abundance. The overall effect of the dam on the smaller mammals, especially several species of mice and woodrats, has been positive. To date, there is no indication that any species of mammal in the river corridor has been either eliminated or reduced in numbers as a result of the changed river regime.

Mice were the first to be noticed as their populations increased in the new habitat. Any river runner with Grand Canyon experience can confirm the accuracy of this statement. The sheer numbers of mice scurrying around most camp kitchens after dark serve to illustrate the success of these small mammals in colonizing the new high-water zone. Around heavily used campsites, the extra food supplied to mice by river runners who spill food into the sand or leave foodstuffs unprotected at night has certainly contributed to the high rodent densities.

Of the eight species of mice known from the river corridor, all have successfully colonized the new high-water zone. The deer mouse is of particular interest, for it was not known from along the river before construction of the dam. Its pre-dam distribution was limited primarily to forest and woodland on or just below the rims.[7] Deer mice were rare in the Inner Canyon, partly because of competition with cactus mice, their ecological equivalent in the open desert at lower elevations and along the river.

The new riparian zone was to prove ideal for the deer mouse. Once the new high-water zone had become established, deer mice evidently reached the river by moving down tributaries whose riparian vegetation provided a more-or-less continuous link between high-elevation woodland and the river. The deer mouse is the common "house" mouse around Page and Flagstaff, Arizona, the very towns from which many river trips originate. It is also possible, then, that deer mice could have been inadvertently introduced into the river corridor in food boxes used by river runners. Whatever their access, deer mice were first beginning to appear in the newly developing high-water zone by 1969. This species later became common in many areas of the river corridor, where it occurred only in and around the dense vegetation of the new riparian zone.[8]

Two other small rodents, the brush mouse and the pinyon mouse, exhibited similar patterns of increase after construction of the dam. The pinyon mouse did not occur in the river corridor before 1963. This species is highly dependent on junipers for nesting sites, and its distribution is almost entirely restricted to pinyon-juniper woodland at and just below the canyon rims. In contrast, the brush mouse was present in the pre-dam river corridor in small numbers. Both the brush and pinyon mouse successfully colonized the new high-water zone after 1963 and increased greatly in numbers.

Ringtails (a relative of the raccoon) and western spotted skunks are observed so frequently at certain river camps that some biologists have suggested that the dam has caused an increase in their numbers. However, these animals quickly adapt to humans and their stores of food. The increased riparian habitat may have contributed to population increases in skunks and ringtails, but it cannot be overlooked that most of the animals seen regularly are merely camp pests.

Definitive population data have never been gathered on the predators of the river corridor, but it is likely that there are more in the riparian zone today than in pre-dam times. Bobcats, coyotes, gray foxes, mountain lions, and raccoons are regularly seen along the river but are very rare. Foxes and coyotes are slightly more common than the cats, but even the former are infrequently seen.

Mule deer and bighorn sheep are commonly seen along some sections of the river, although there is no information indicating whether or not the dam has influenced them in any way. Deer occur in higher densities in the woodland and forest of the rim country, moving into the river corridor in larger numbers mostly during severe winters when deep snowpack on the rim drives them downslope. Bighorn frequent the rugged side canyons away from the river, but they also come to the river to drink and feed during the hottest parts of the summer.

Lizards, Snakes, and Toads

Amphibians and reptiles seem to have benefited greatly from the new river regime. Thousands of Rocky Mountain and red-spotted toads, for example, invade some river camps at night. These toads can be so abundant in the new high-water zone on summer nights that it is almost impossible to walk through camp without stepping on one. During the height of the toads' breeding activities, many beaches are covered with the tracks of nightly toad wanderings from the desert and riparian areas to the river's margin. Unlike most amphibians, toads are primarily terrestrial, living in deep burrows during the day and venturing out under safety of darkness to breed and lay their eggs in the water.

Other amphibians found in the river corridor include a single record each of the tiger salamander and leopard frog and occasional appearances of the spade-footed toad. Common in the tributaries, but almost never found along the mainstem river, is the tree frog. This noisy frog, never exceeding a silver dollar in size, has a loud breeding call that is identical to the "baa"

Figure 7.2. Mule deer are frequently seen along the river, especially near Saddle Canyon and downstream of Lava Falls. As is the case for most of the larger riparian mammals, the effect of the changing river environment on deer is unknown. *Dugald Bremner*

sound of sheep. A moonlight walk up a side canyon can be a nerve-wrenching experience when these little frogs unexpectedly call from a few feet away.

Amphibians and reptiles have benefited by taking advantage of the new resources provided by the dam in a slightly different way than the birds and mammals. Birds and mammals concentrate mainly on the riparian vegetation of the new high-water zone for cover and secondarily on its abundant insects and other food resources. Amphibians and reptiles, on the other hand, focus primarily on the new insect resources and only secondarily, if at all, on the new vegetation. This pattern is indicated by the trend for lizards, in particular, to be most abundant at the water's edge and to become less abundant with increasing distance from the river, regardless of the type of vegetation present.[9]

Rattlesnakes are among the most common of the larger reptiles found along the river. The new high-water zone is the habitat apparently favored most by this fearsome predator. Two species of rattlers are common: the more abundant pink-colored Grand Canyon rattlesnake and the less colorful Mitchell's rattlesnake. Both of these snakes are quite docile and rarely pose more of a problem than simply scaring the wits out of hikers and river runners. The black-tailed rattlesnake is found only infrequently and is much larger than the other two, sometimes reaching 5 feet in length. This snake is often aggressive, and encountering one unexpectedly at close range can be an unpleasant experience that will not soon be forgotten.

Other regularly seen snakes of the river corridor include coachwhip, California kingsnake, gopher snake, ring-neck snake, and ringed ground snake. No information on snake densities by habitat type exists, and it can only be inferred that the dam has had an impact on their abundance. Since most snakes prey on small mammals and other reptiles, both of which have increased in the new high-water zone, foraging conditions for snakes have markedly improved since the dam.

The common lizards of the river corridor, in order of decreasing relative density, are side-blotched, western whiptail, desert spiny, and tree lizard. The tree lizard is least common in the riparian zone, because in spite of its name, it almost never occurs in vegetation but prefers the sheer rock walls near the water's edge. The other three common species, however, are found in greater density in the new high-water zone along the Colorado River than in any other habitats thus far studied in the Southwest.[10]

In the new high-water zone, lizard densities are sometimes up to ten times higher than in adjacent desertscrub or in the old high-water zone. The species are the same, but the numbers change dramatically. This finding is surprising when one considers that probably few, if any, lizards occurred in the scour area of the pre-dam shoreline. Not only are there more lizards found today in the riverine environment than elsewhere throughout the West, but these densities are exhibited in a habitat that did not exist before Glen Canyon Dam.

The growth of the lizard population is primarily a response to the increased availability of food in the new high-water zone. Like the insect-eating birds and bats, the lizards take full advantage of the new productivity of the river. The insects hatching and emerging from the river often swarm into the new high-water zone, landing on vegetation as well as rocks and soil, where they are easy prey for lizards. One peculiarity of the new system is that some of the lizards have even learned to forage in the intertidal area produced by the daily high and low river flows. The western whiptail is the champion of the intertidal zone, where it can be found following the dropping water's edge, feeding on harvester ants, stranded amphipods, and black flies.

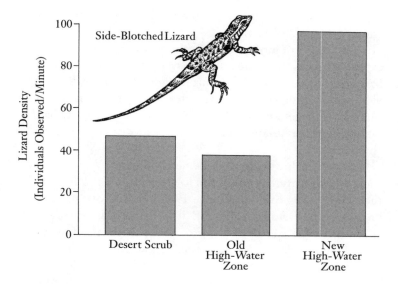

Figure 7.3. Lizards are most abundant in the new high-water zone, especially at the water's edge, where they have learned to forage for insects and amphipods in the intertidal zone.

Riparian Wildlife and Post-Dam Flooding

The increasing density and diversity of wildlife in the new riparian zone was brought to a temporary halt by the flood of 1983. The same floodwaters that scoured away much of the vegetation zone also drowned or displaced most amphibians and reptiles below the 92,000-cfs water line, swept away mammals unable to escape the rising water, and inundated scores of active birds' nests. Only the effects of the flood on bird populations have been well documented, and these data must serve as a generalized example of the impact of post-dam flooding on riparian wildlife.

The timing of the 1983 flood was particularly bad for nesting birds. Their nesting season is limited to that brief period each year when conditions are just right to provide adult birds and their young with sufficient food, warmth, and cover. For those species that nest in the river corridor, the season may last from March to July. The all-important peak of nesting, when the majority of birds of all species are occupied with the raising of young, is during the last two weeks of May and the first two weeks of June.

The initial rise of the flood to near 50,000 cfs in early June 1983 caught most birds at the peak of the nesting season, with eggs or young birds still helpless in the nest. The immediate impact of the flood was not on adult birds, because they could easily fly above the rising water, but on that year's reproductive effort, the eggs and nestlings.

Most species of birds nesting in the new high-water zone experienced nest loss when the vegetation was scoured away or bent over by high water. Not surprisingly, those species habitually nesting closest to the water and lowest to the ground suffered the greatest losses. Three species were especially hard hit: the common yellowthroat, Bell's vireo, and yellow-breasted chat.

Common yellowthroats usually nest less than 2 feet above the surface of the ground or water in low, marshy areas. Most of their nests in the river corridor in 1983 were lost because of flooding. The average height of Bell's vireo nests is also low, at approximately 3 feet above the ground. More than half of their nests were inundated and destroyed. Yellow-breasted chats nest just slightly higher in the vegetation than Bell's vireo, on the average, and chats experienced an approximately 20 percent loss of their nests.[11]

Most birds have adapted to occasional losses of their nests by developing the ability to build a new nest quickly and lay a new clutch of eggs if necessary. After the initial rise of the floodwaters to about the 50,000-cfs level, many species whose nests had been inundated were soon able to renest. This strategy would have worked, except that the floodwaters kept rising, stairstep-fashion, to more than 90,000 cfs and sequentially inundated most renesting attempts. Some pairs of adult birds were known to have built three or more nests, all of which were eventually lost to the flood. Furthermore, the flood continued throughout the remainder of the nesting season and prevented some species from successfully nesting at all in 1983.

The main impact of the flood on birds in the short-term was the loss of a part or all of that year's reproductive efforts. One long-term effect of the flood, however, was eventually to cause a decline in bird population numbers. This decline would not become apparent until 1984, it was thought, when the adults of those species that were hardest hit by the flood again returned to the river corridor to nest. The rigors of migration to and from Mexico would take their toll of adult birds over the winter, and fewer young birds had been produced in the 1983 nesting season to replace those adults that would be lost from natural causes.

As predicted, common yellowthroat and Bell's vireo populations had declined precipitously by the spring of 1984. Bell's vireo has since recovered, as have most other species of birds nesting along the river, even though the vegetation has not. In comparison, common yellowthroats continued to decline

Figure 7.4. Common yellowthroat, Bell's vireo, and yellow-breasted chat experienced substantial nest losses because of inundation of their nesting habitat of riparian vegetation by rapidly rising floodwaters in June 1983. Future high water releases at the peak of the nesting season could cause similar losses, although the level of riparian vegetation relative to water levels has changed somewhat since 1983.

through 1987 as a result of the almost complete loss of their preferred habitat of cattail marshes. Recovery of the common yellowthroat population will probably not occur until cattail marshes reestablish themselves in the river corridor.[12] The flood did favor the widespread establishment of cattail marshes on the mud flats of the upper Lake Mead area, marshes that were no doubt colonized by common yellowthroats soon thereafter. This unexpected turn of events on the lake may eventually help yellowthroats to recolonize the remainder of the river corridor and recover in numbers.

The most important long-term effect of the 1983 flood on riparian nesting birds, as well as many other species of riparian wildlife, is indirect. The extent and structure of vegetation in the river corridor is what really dictates the abundance and status of bird populations, so the long-term effect of the flood on plants in the new high-water zone will, in essence, control the size of future bird populations. If the vegetation of the new high-water zone can recover in both extent and composition from the losses caused by the flood, bird and other wildlife populations will recover as well. The river's flow regimes, sediment balance, and nutrient levels, all of which play a role in the development of riparian vegetation, indirectly influence birds to a great extent.

Lizard populations at the water's edge were also dramatically reduced by the flood, although this decrease is not well

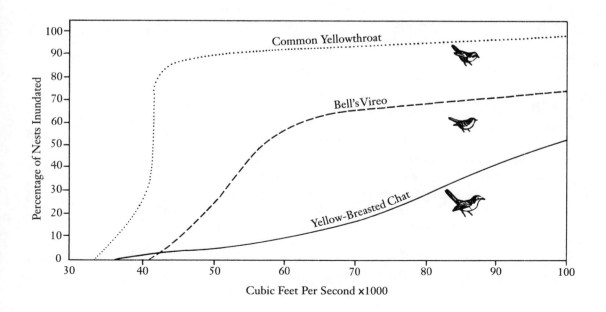

documented. The decline of lizard populations during and after
the flood was felt immediately because adult and juvenile lizards
alike were destroyed or washed away. Lizard numbers were
extremely high at the water's edge by 1984, though, showing
that rapid colonization from the old high-water zone and desert
had taken place within a short time. In the most dramatic
instance, an inundated cobble bar more than 300 feet from the
nearest area that had not been inundated by the flood was noted
to exhibit high lizard densities by the spring of 1984. This case
illustrates both the rate and distance in which lizards are able to
disperse within a year.[13] Such animal species are said to be
resilient, a term indicating their ability to recover rapidly from
catastrophe.

Wildlife Along the Ancient Colorado River

Dozens of animal species have alternately colonized, occupied,
and then disappeared from the river corridor over the last thirty
thousand years. These constant changes in the natural
distribution of animals along the river were primarily the result
of widespread climatic changes over western North America
which gradually transformed the vegetation of the entire
Southwest.

The most recent climatic change that strongly altered plant
and animal life along the river was at the end of the last ice age,
some eleven thousand years ago. Previously cool and wet, the
climate shifted to warmer, drier conditions at that time and, in
combination with other factors, caused many species to leave the
river corridor or become extinct. These changes, however,
occurred gradually over a period of thousands of years. In
contrast, the rate of change in animal populations caused by
Glen Canyon Dam was rapid. Thus changes in the species of
wildlife along the river are nothing new, but the rate at which
modifications have occurred as a result of the dam is
unprecedented.

On a raft trip down the river of twenty thousand years ago, if
such a thing were possible, twentieth-century humans would see
wildlife both familiar and unfamiliar to them. Mule deer,
bighorn sheep, gray fox, coyote, ringtail, and raccoon were all
present along the river then. The rafters would be surprised,
however, to see animals extinct today, such as the Shasta ground
sloth, camel, a wild horselike animal, and Harrington's mountain
goat, living along the river thousands of years ago. In the skies
above the river of that time were familiar birds such as the
turkey vulture, red-tailed hawk, and canyon wren. Also present

Table 7.2 Animals of the River Corridor 10,000 to 30,000 Years Ago but Now Absent

Extinct Species	
Birds	California condor
	Another extinct species of condor
	Merriam's terratorn
	An extinct species of wild turkey
Mammals	Harrington's mountain goat
	Camel
	A horse or burrolike mammal
	Shasta ground sloth
Absent Species	
That Occur Nearby	
Birds	Common black-hawk
	Sage grouse
	Barn owl
	Northern shrike
Mammals	Desert shrew
	Desert cottontail rabbit
	Black-tailed jackrabbit
	Porcupine
	Pronghorn
Reptiles	Desert tortoise
Absent Species Occurring in Other	
Parts of North America or Mexico	
Birds	Aplomado falcon
	Clay-colored robin
Mammals	Yellow-bellied marmot
	Bison

were the California condor, as well as another species of now-extinct condor, and a huge scavenging bird with a 12-foot wingspan, named Merriam's terratorn.

California condors almost persisted through the climatic and habitat changes since the ice age to become inhabitants of the post-dam river corridor. The last California condor ever seen in the Grand Canyon was shot in 1881 by prospectors along the Colorado River in the vicinity of Pearce Ferry, now on Lake Mead. This species is now absent from the canyon and almost extinct in North America. In 1987 the last of the wild condors, some twenty-five individuals in California, were live-captured and placed in zoos in a last-ditch effort to save the species. Half of them were placed in the San Diego Zoo, the other half in the Los Angeles Zoo.

Condors began declining in numbers in the last half of the nineteenth century for a variety of reasons. The overwhelming threat, which led to the capture of the last of the wild birds in California, was lead poisoning. Condors feed on carrion, and most of the dead animals available to them by the 1980s were contaminated by lead — bullets lodged in the carcasses of deer that had been wounded by hunters. Lead poisoning had become the primary cause of premature death in these magnificent birds. The hope now is that enough young condors can be raised in captivity for future return to the wild. In the zoos dedicated biologists have succeeded in raising healthy young, and releases of condors back into the wild may soon be a reality.

One of the remaining problems facing the endangered condor is an ideal locality for its eventual reintroduction. The remote open space of the Grand Canyon and vicinity contains some of the best habitat left in North America where the condor would be guaranteed a fresh start. The congested, contaminated areas of southern and central California, where the last adults were captured, still exhibit the environmental hazards that drove the species to near-extinction. To reintroduce the condors to these areas would be almost counterproductive. Grand Canyon has no such hazards, and the condor could be reintroduced to the safety of its skies with a higher probability of success.

Evidence for the prehistoric existence of condors and other ancient inhabitants of the river corridor comes from collections of animal bones preserved in dry caves along the river. Bone caches have provided a detailed chronology of animals present between ten thousand and thirty thousand years ago. Stanton's Cave has furnished the most animal remains,[14] and identifiable animal bones have also been collected in Rampart Cave and other caves of the river corridor.[15] Animal remains as delicate as eggshells and dried skin have been perfectly preserved in the dry atmosphere of the caves. Radiocarbon dating techniques can be used to indicate when the animals were alive.

Analysis of these remains has revealed that most of the animals present along the river today were also present ten thousand to thirty thousand years ago. Perhaps more interesting, though, are those prehistoric animals that no longer occur in the river corridor.

At least one group of animals that formerly occurred along the river provides an intriguing mystery. Rabbits are one of the most common and widespread groups of mammals in North America today, occupying virtually all habitats from alpine tundra to arid desert to dense swamp. Yet no rabbits exist along the river corridor of today. Bones from Stanton's Cave have established that both the black-tailed jackrabbit and the desert cottontail rabbit were present thousands of years ago. The cause of their demise is uncertain, but it may have been related to long-term climatic and vegetation change. Interestingly, rabbits are common in adjacent areas outside the Grand Canyon region at elevations and in habitats similar to those found along today's river. Rabbits have evidently failed to recolonize the river corridor, and their absence is one of the enigmas of the distribution of animal life in the Inner Canyon.

One species of mammal that became extinct after the ice age had the dubious distinction of being temporarily replaced by a similar, introduced species in recent times, A species of wild horse, or burrolike animal, occurred in the river corridor between ten thousand and thirty thousand years ago. Evidence consists of the discovery of part of a backbone and part of a hoof found in Rampart and Stanton's caves. This "horse" became extinct ten thousand years ago, but its ecological equivalent, the burro, was introduced into the canyon by prospectors in the late 1800s.

Feral burros were commonplace along the river by the 1970s, and their impact on soils, vegetation, and native wildlife was severe.[16] The sporadic control programs that had been implemented by the National Park Service since the 1920s effectively kept the growing feral burro herds in check, but these activities were outlawed by Congress in the late 1970s. The solution was for the feral burros to be removed by humane methods. The Fund for Animals, Inc., rounded up and airlifted all feral burros from the canyon and the river corridor in 1981 at a cost of nearly a million dollars. Feral burros are now absent from the river corridor, except for the individuals that occasionally stray in from the adjacent Lake Mead National Recreation Area.

The River as a Barrier

Although the dam, man, and the changing climate of prehistoric times have greatly modified many features of animal life in the river corridor, other aspects of the distribution of wildlife remain as they have for centuries. A large river and a deep canyon, with or without a dam, pose an overwhelming obstacle to the dispersal of some land animals. Such animals include those that are unable to fly or swim, such as lizards, rodents, and other small mammals. Unlike the situation in the riparian plant community, where the river serves as a connection between the hot southern deserts and cold northern desert, the river acts as a barrier to some wildlife. The result is that many smaller animals are effectively restricted in distribution to one side of the river.

The most dramatic example of the river as a barrier to animal distribution is provided by two small species of mice. The long-tailed pocket mouse and the rock pocket mouse are both common in the river corridor and occupy similar habitats dominated by rocky soil or rock outcrops. For all their similarities, these two species are separated by one important feature: the Colorado River.[17] The long-tailed pocket mouse is found only on the north, or right, bank of the river; the rock pocket mouse occurs only on the left, or south, bank.

A similar example is provided by a species of snake occurring in the river corridor. The gopher snake is found on both sides of the river, yet a different subspecies, or variety, is restricted to each side.[18] The Great Basin gopher snake exists only on the north bank of the river. The Sonoran gopher snake, differing mainly in having more dark spots on its back, is found only on the south side of the river.

Most snakes can swim better than mice, enhancing their chances of colonizing the opposite shore of the river. Although not a total barrier in the case of the gopher snake, the river does greatly slow down the movement of individuals from one side to another. Thus the reproductive contribution of any colonizing individuals to the overall pool of inherited traits of the species on the other side of the river is reduced. The result of limited across-river movement is for the populations on each side to behave as if they were separate. As a partial barrier, the river tends to favor the evolution of one subspecies or variety with a specific set of inherited traits on one side of the river and another on the other side.

The presence of different subspecies or varieties of animals on opposite sides of the river is more common than the presence of different species on opposite banks. The more mobile the animal and the better it can swim, the less difference exists between varieties on the north and south banks. Until as recently as the 1960s, the river corridor was seen as a barrier to the distribution of more than ten sets of subspecies.[19] Recent advances in taxonomy, the science of classification of organisms, have indicated fewer real instances of the river as a barrier.

In addition to the two subspecies of gopher snake, examples of species with different subspecies more or less on opposite sides of the river include the northern grasshopper mouse and the mountain lion.[20] Many other species and subspecies are separated more by the canyon itself than by the river. The

Abert's and Kaibab squirrels on the South and North rims, respectively, are the most notable example.

Life at the Top of the Food Chain

The substantive modifications caused by the dam in water flow, sediment budget, and aquatic productivity of the river have had profound effects on the riverine food web, or chain. Sometimes these relationships are relatively easy to follow, as in the case of the simple food chain of trout and eagles at Nankoweap Creek. Other examples of food web cause-and-effect relationships are more complex.

Perhaps the most interesting hypothesis addressing potential dam-related changes in wildlife populations concerns the endangered peregrine falcon. Studies in the late 1980s indicated that Grand Canyon National Park supported the largest known population of breeding peregrine falcons in the contiguous forty-eight states. The majority of the falcon population is known to nest along the river corridor, where high densities occur from Lees Ferry to Lake Mead.[21]

Unfortunately, it is impossible to determine what numbers of falcons were present before the dam. Nevertheless, a strong circumstantial case can be made that the changes brought by the dam have benefited peregrine falcons and may be partially responsible for their abundance in the river corridor. The difficulty of scientifically testing this hypothesis is overwhelming, but the hypothesis itself is useful because it illustrates the subtle complexity of post-dam changes moving up the riverine food chain.

In essence, the increased primary productivity of the river may have ultimately allowed the peregrine falcon population to expand. The story begins low on the food chain within the river, where untold billions of aquatic insect larvae mature and emerge. It is believed that the species of midges, flies, and gnats resulting from the changed ecosystem were not common before the dam. The presence of these insects each spring and summer serves as a significant supplemental food source to insectivorous riparian wildlife — bats, swifts, and swallows — that constitutes a substantial portion of the prey base of adult falcons. Historical data are lacking that would clearly substantiate increases in these insectivorous birds, but the abundance of post-dam bats, swifts, and swallows suggests that they have increased in numbers because of dam-related changes in the riverine ecosystem. If this scenario is accurate, then it is only a small move up the food

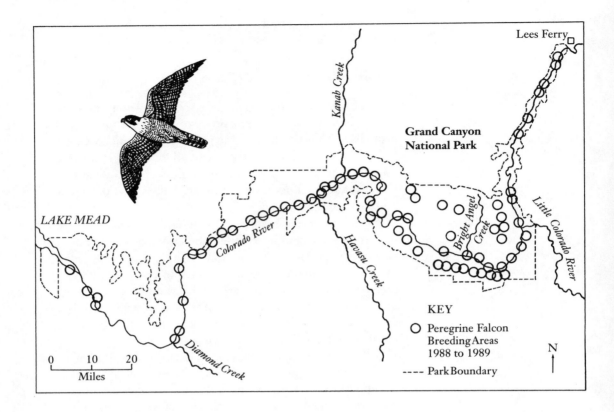

LAKE MEAD

Colorado River

Diamond Creek

0   10   20
Miles

Kanab Creek

Havasu Creek

Grand Canyon National Park

Bright Angel Creek

Little Colorado River

Lees Ferry

KEY

O Peregrine Falcon Breeding Areas 1988 to 1989

---- Park Boundary

N

chain to the peregrine falcon, the top predator. The extra prey provided indirectly by the dam would be more than enough to support an increase in the peregrine falcon population.

Additional factors may also influence the situation. Migrant waterfowl, the mainstay of most peregrine diets in late winter and early spring, may have increased in abundance along the river corridor because of the overall growth of aquatic productivity. If so, the additional prey could provide more food to the falcons at a key time when more energy is needed for the laying of eggs and other reproductive activities.

The hypothesized relationship between the dam and increases in peregrines is further complicated by the fact that falcon density is growing nationwide as a direct result of the decreased use of some highly toxic pesticides since the 1960s. Pesticides used to control insect pests on crops were also assimilated into the bodies of insects not directly killed. These insects were preyed on by insectivorous birds such as swifts and swallows, which in turn were preyed on by top predators like peregrine falcons. The pesticides became more and more concentrated with each step up the food chain, resulting in dangerous accumulations in the tissues of falcons.

High levels of chlorinated hydrocarbons like DDT in the tissues of birds of prey cause eggshell thinning, and peregrine eggshells were especially affected. Widespread nest failure and the eventual decline in population density brought peregrine falcons to the brink of extinction across most of North America by the late 1960s.

The documented increase in peregrine falcon abundance in the 1980s due to recovery from the effects of DDT may be acting hand in glove with the increase in prey caused by the dam, together resulting in high peregrine densities along the river. Unfortunately, there will never be any way to ascertain the historic abundance of falcons or to separate the positive influences of the dam from the decrease in pesticide use on the falcon population in Grand Canyon.

The indirect effects of the dam on nesting peregrine falcons and the overall increase in riparian animals are the best examples of how the new river has beneficially influenced wildlife. In contrast, the effects of the 1983 flood show how the pendulum of dam-caused changes can swing the other way. The dam is now the greatest single influence over the wildlife of the river corridor and needs to be recognized as such. If the beneficial changes in wildlife brought about by the dam are to be maintained, modifications must be made in how the dam is operated.

Because of the general lack of natural resource goals in the operation of the dam, animal populations of the river corridor are unstable and subject to wild swings in abundance. Only by careful, directed management of water releases will it be possible to bring the animals of the river corridor toward a balanced, dynamic equilibrium in the future. And it is a virtual certainty that any changes in the management of water releases from the dam will be somewhat at odds with the deeply entrenched political and economic factors that influence how the dam is operated.

Figure 7.5. Grand Canyon National Park exhibits the largest population of nesting peregrine falcons on a single land-management unit in the United States outside Alaska. Most nesting pairs are found along the Colorado River, where they occur an average of 4.2 river miles apart during spring and summer. Some pairs may be separated by fewer than 2 river miles.

PART IV   The Biopolitical Environment

8 BIOPOLITICAL MANAGEMENT OF THE RIVER

The rafters were nine days out on an eleven-day river trip when they ran into trouble. The water must have been running at least 22,000 cfs when they pulled into the lunch stop just below Deer Creek, and with the high water they expected to make the 30 miles to the National Canyon camp easily by dark. The boatman was well aware that weekend low water, released from the dam two days previously, was on its way downriver, but the lower flows were not expected until the following day. After quickly eating, they decided to take a short hike into the upper gorge of beautiful Deer Creek Canyon. Making sure the 33-foot pontoon boat was securely tied to a tamarisk on the beach, the boatman joined his passengers for the pleasant hike. The water must have begun dropping the moment he turned his back.

As the hikers worked their way deep into the chasm of Deer Creek, splashing through crystal pools and reveling in the wildness of this unspoiled canyon, they relaxed in the knowledge that they were far from their commonplace worlds. The last thing on their minds was that they would be affected by an aberration in the western power grid.

When the hikers returned two hours later, they were dismayed to find their five-ton boat high and dry, 30 feet from the river's edge. The rafters would spend that night and possibly the next on the narrow beach where their boat was stuck in the sand. Experienced river guides take pride in their ability to predict water-level fluctuations below Glen Canyon Dam. Knowing what the water will do prevents unpleasant incidents like smashing motors on barely submerged rocks or beaching boats. Once a heavy motorboat is stranded, the group must either wait for the next cycle of high water or hope a passing river party will help push the boat back into the water. About forty strong people are required to push a fully loaded 33-foot boat into the water.

A Question of Balance

The drastic change in water level experienced by the Deer Creek party is typical of the water fluctuations caused by the operation of Glen Canyon Dam. Water releases from the dam are governed by an array of legal, political, economic, and climatic constraints. Decisions made from Los Angeles to Washington, D.C., from as long ago as 1922 to as recently as yesterday's federal district court decision, and by dozens of agencies, organizations, and institutions influence the dam's operation.

Glaring contradictions in policy and mandate among the different agencies and special interest groups dictate that there is no universally accepted set of goals bonding the diverse criteria influencing the dam. Most importantly, the overwhelmingly complex set of agencies, organizations, and mandates ensures that there will be no coordinated management of the dam and the river.

The result has been the steady degradation of both the natural resources and the wilderness experience along the river in Grand Canyon. Even a slight change in the focus or priorities of only one of the powerful forces influencing operation of the dam can have far-reaching consequences, capable of setting the entire river ecosystem out of balance. The natural systems of the river will be unstable until the agencies controlling the operation of the dam develop and abide by a set of common management goals.

The Reason for Glen Canyon Dam

The Colorado River Compact of 1922 began a slow process that eventually led to the construction of Glen Canyon Dam. The compact accomplished what many considered to be impossible — the formulation of a plan by the seven states comprising the Colorado River basin to divide the river's water among themselves.[1] The compact was unique, for it was the first time a group of states agreed to divide the water of an interstate stream for consumption.

For the period of record from 1906 to 1922, annual natural flow of the river averaged a little over 15 maf. The upper basin states upstream of Lees Ferry (Colorado, Wyoming, Utah, and New Mexico) guaranteed delivery of 75 maf each ten years (an average of 7.5 maf each year, or approximately half the river's annual flow) to the lower basin states (Arizona, Nevada, and California). Although most of the states agreed in principle to the water distribution, Arizona felt its share was insufficient and chose not to ratify the compact for many years.

The seven states neglected to allocate any water for Mexico, creating an international tension that was finally relieved by the Mexican Water Treaty of 1944. The treaty allowed Mexico 1.5 maf per year, an amount that was to be contributed equally by the upper and lower basins.[2] The final amount of water that the upper basin must provide the lower basin (7.5 maf per year), including Mexico (0.73 maf per year), comes to 8.23 maf annually. The original allocation of water, however, was based on an unusually wet hydrological cycle. In the long-term perspective, especially if a severe drought is experienced, more water has been allocated than exists.

Construction of a dam near Page, Arizona, with the primary purpose of regulating the delivery of water from the upper to the lower basin, was authorized by the Colorado River Storage Project Act of 1956. The 1956 act designated Glen Canyon Dam as the regulatory spigot between the basins.

After water-delivery mandates, other purposes of the dam included water storage and flood control, and the generation of hydroelectric power as an incidental purpose.[3] Recreational and environmental concerns were to be addressed by the operation of the dam, but they were not specifically identified as primary goals.

The Colorado River Basin Act of 1968 further defined the hierarchy of priorities for Glen Canyon Dam. The act reaffirmed the mandate of water delivery first, followed by flood control, water storage, environmental and recreational concerns, and power generation as an incidental objective. These two acts and the compact form the framework of the "law of the river." Court cases, judicial interpretation, and treaties have further refined the legal requirements that have resulted in the present management system.[4]

With these three powerful pieces of legislation solidly in place, the number of other factors directly influencing the operation of the dam quickly began to multiply. This proliferation of interests soon resulted in a heightened level of complexity surrounding decisions affecting dam operations. The Bureau of Reclamation, a bureau within the U.S. Department of the Interior, was originally charged with operating and maintaining the dam and marketing its hydropower. In 1977 the Western Area Power Administration (WAPA) of the Department of Energy assumed responsibility for marketing and selling the hydropower under the Department of Energy Organization Act.[5]

Much to the dismay of the water and power interests, the enactment of two major environmental laws, the National Environmental Policy Act (NEPA) of 1969 and the Endangered Species Act of 1973, demanded a consideration of how the operation of the dam might affect the downstream riverine ecosystem. And finally, notwithstanding WAPA's array of computers and energy contracts and the Bureau of Reclamation's operating plans, Mother Nature still has some veto power. Annual precipitation in the Rocky Mountains can take precedence over all of these interests, except for the minimum water-delivery provisions of the 1922 compact. Unusually wet years may see more water passing downstream, with higher, more steady flows. For example, in 1984, 21 maf of water was released. In drought years 8.23 maf is still released from the dam, but in a manner that results in lower overall flows marked by extreme fluctuations.

Other groups indirectly affect the dam's operation. For example, the National Park Service, originally unconcerned about the effect of the dam on downstream portions of Grand Canyon National Park, now manages natural resources and "wilderness" recreation in a river corridor that is no longer pristine and uninfluenced by the hand of man.[6] River recreationists, in turn, lobby their congressional representatives in hopes of modifying dam operations to obtain favorable water levels for recreation. Most river runners prefer water levels that are neither too high nor too low; a level between 10,000 and 20,000 cfs makes for ideal rafting conditions. Changing water levels are unpopular with rafters, because the fluctuations detract from the perception of a wilderness experience and also sometimes leave boats stranded.

The U.S. Fish and Wildlife Service indirectly regulates the dam by interpreting the effects of its operation on endangered species in Grand Canyon. The Colorado River Electrical Distributors Association (CREDA) exerts a tremendous amount of influence over the Bureau of Reclamation and WAPA, influence that has helped to dictate the schedule on which water is released from the dam. Electric power districts, irrigation districts, water conservation districts, fishing enthusiasts, river runners, and environmentalists all attempt to bend dam operations to benefit their special interests.

These diverse political elements operate without a common mandate or common perspective on how the dam is to be managed. The mandate of the Bureau of Reclamation is water delivery; WAPA's purpose is to sell electricity; CREDA strives to maintain maximum availability of inexpensive power; National

Park Service management goals are split between the preservation of natural values and the need to provide for the public enjoyment of those values (with the latter goal having political priority)[7]; the U.S. Fish and Wildlife Service considers the needs of migratory and endangered species; and the Arizona Game and Fish Department considers the needs of native wildlife, although their primary concern is the sport-fishery. The result is that each group promotes its own agenda and special interests. Not only are the agencies prevented from seeing eye to eye by their different management philosophies and mandates, they often do not even acknowledge one another's existence.

Overall river and dam management that will balance the various interests is the responsibility of the secretary of the interior as the legally designated water master of the Colorado River. The principal reason such a plan has never been developed is that the water and power interests within the Departments of the Interior and Energy have successfully dominated river management, deliberately preventing a more holistic interpretation of existing law.

The Water Masters

The U.S. Department of the Interior oversees the management of the river as defined by the 1922 Colorado River Compact and the other laws pertaining to the Colorado River. In turn, the secretary of the interior charged the Bureau of Reclamation with managing the dams and reservoirs, including responsibility for meeting the water delivery requirements, and generating electricity.

The Bureau of Reclamation has four major criteria for determining the amount of water released from Glen Canyon Dam. The first criterion is to ensure the *annual* release of 8.23 maf. Not a drop of water more, if possible, and certainly not a drop less flows through the dam in any given year.

Before the filling of Lake Powell in 1980, annual inflow to the lake above the 8.23 maf minimum was retained behind the dam in order to reach maximum reservoir capacity as quickly as possible. Now that the reservoir has filled, however, wet year runoff occasionally requires the release of more than 8.23 maf. If the additional water discharges exceed the capacity of the generators (33,200 cfs), the excess is routed through the outlet tubes and the two spillways. Such releases have the potential to cause downstream flooding and resource degradation such as occurred from 1983 to 1986. But the Bureau of Reclamation (as

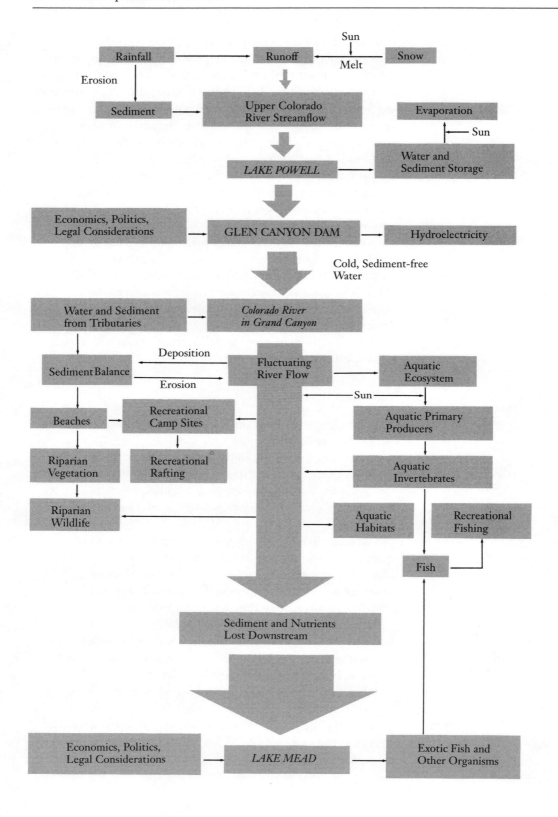

well as WAPA and CREDA) is loath to allow excess water to bypass the dam without running through the power plant, because the release of excess water through the river outlet tubes or spillways represents lost hydropower revenues and "wasted" water.

The second major criterion determining how the bureau releases water concerns the *monthly* release schedule. The monthly volumes are primarily based on meeting water commitments and avoiding spills. An annual operating plan, based on a water year that extends from October through the following September, is initially developed by the bureau for the entire river basin. This plan is constantly fine-tuned as the water arrives. The dam's annual operating goal is collectively met by a series of planned monthly water releases that total 8.23 maf. The overall volume of water to be released each month may range from approximately 200,000 to more than 600,000 acre-feet, depending on season, lake level, upper basin snowpack, and river inflow to Lake Powell. Long-term weather forecasts performed by the National Weather Service and snowpack estimates provided by the Soil Conservation Service are also a key ingredient in arriving at the correct monthly release formulas.

The months with higher or lower release volumes are generally predictable. Typically, in June and July there is high river inflow to Lake Powell, as well as high demand for hydroelectric power, so these months tend to exhibit larger release volumes. Spring and autumn months are usually characterized by lower inflow to Lake Powell and lower power demands and generally exhibit lower release volumes. Some flexibility is built into the monthly release system to account for variable weather conditions. But whatever the monthly releases, the Bureau of Reclamation strives to release 8.23 maf annually and to have the reservoir at 22 maf on January 1 of each year and full (27 maf) by July 1. Because of the consequences of the 1983–84 high flows that forced spillway releases, the bureau now manages the lake to stay approximately 5 feet below full capacity to accommodate unexpected inflow.

The third criterion concerns the upper limit of water that can be released at any given *instant*. The Bureau of Reclamation is administratively limited to releasing a maximum of 31,500 cfs, even though the full capacity of the generators is 33,200 cfs. This discrepancy between actual and potential maximum releases will continue to exist until completion of the Glen Canyon Environmental Studies.[8] Surplus water releases above 33,200 cfs which are necessary in times of emergency, such as during the 1983 flood, are not affected.

Figure 8.1. The flow of the Colorado River through Grand Canyon is influenced by, and in turn influences, an overwhelming number of climatic, economic, environmental, legal, political, and recreational factors. This simplified flowchart illustrates some of the factors and relationships that come into play upstream of Lake Mead.

The fourth criterion is a nonbinding, informal agreement between the Bureau of Reclamation and the National Park Service, acting as representative of the recreational community. The agreement calls for the bureau to maintain a 1,000-cfs minimum release during the nonrecreation season (Labor Day to Easter Sunday) and a 3,000-cfs minimum release during times of peak recreational use of the river. In theory, the conditions of this agreement provide a limited guarantee of reasonable instream flows for recreation and, as an apparent afterthought, the protection of environmental values. In reality, the agreement is regularly violated because power demands override all other considerations.

The Power Brokers

Once the Bureau of Reclamation arrives at the monthly volumes of water to be released, most of the control over how and when the water is released passes to WAPA. Daily, hourly, and minute-by-minute releases are solely the responsibility of the power brokers. The basis for their decisions is a complex set of guidelines and contracts directed entirely by economic factors relating to the sale of hydroelectric power, including long-term hydroelectric power contracts, power demands, and fluctuating power prices. Except during emergencies, WAPA controls the water releases by telephone and computer, long-distance from Montrose, Colorado.

WAPA was empowered by the 1977 Department of Energy Organization Act to market and transmit federally produced electric power in the western states. In turn, WAPA collects power revenues that are allocated to repay construction, maintenance, and operating costs of its power-generation facilities, participating irrigation projects, and other expenses of the Colorado River Storage Project (CRSP). WAPA oversees the distribution of more than forty billion kilowatt-hours of electricity each year from California to Texas, enough power to serve more than twenty million people.[9]

Kilowatt-hours that can be generated within an instant's notice at Glen Canyon Dam are of special value. Electric power is worth more seasonally and during daylight hours on weekdays because of greater electrical demand at those times. The result of this daily and hourly fluctuation in electric power value is vividly reflected in the daily and hourly fluctuations of water levels surging through the Grand Canyon. More water is released during the day and less at night, resulting in the daily rising and falling of the "tide" along the river.

Flexibility is the key to earning hydropower revenues from Glen Canyon Dam. And hydropower is of more value because of its unique flexibility in the WAPA power network. The reason is simple and has to do with the ability of hydropower-generating facilities to meet peak demands quickly and easily. For example, electric power generators at the Navajo Generating Station, a coal-fired facility near Page, Arizona, take twenty-four hours to reach full generating capacity once they begin operation. In contrast, the generators at Glen Canyon Dam take only minutes to reach full generating capacity once the turbines begin to spin. Glen Canyon Dam can cope with distant power demands almost instantly and, as a result, has the potential to bring in more power revenues from the sale of electricity.[10]

The sequence of events that eventually determines how much water is released at any given time begins in WAPA's administrative offices in Golden, Colorado. The all-important power contracts with electrical power consumer groups are primarily developed here, as well as in the area office at Salt Lake City, Utah. The Salt Lake City area office develops a general plan of operation based on the monthly release volumes supplied by the Bureau of Reclamation, taking into consideration the firm contract requirements and the power market. Power brokers in Montrose are on the telephone around the clock in a never-ending race against time, in order to meet contract delivery schedules and to get the highest price for every last kilowatt-hour of electricity that can be squeezed from the monthly release volumes.

Flexibility has been carefully built into every step of the power production process, save one. Oddly enough, the power contracts between WAPA and the consumer groups are rigid, long-term economic commitments that essentially demand selectively maximized power generation. Power contracts of up to twenty years' duration were not uncommon once the dam began producing hydroelectricity. Contracts of this length required WAPA to supply guaranteed amounts of power and utilities to purchase a minimum amount over a period of time with an incalculable number of unforeseen variables.

Much of the uncertainty inherent in long-term power contracts of this nature has been eliminated by maximizing power releases and minimizing the other mandates for the dam. With the shift of control of power generation and marketing from the Department of the Interior to the Department of Energy in 1977, there has been a corresponding shift in the implementation of the provisions of the 1956 and 1968 Colorado River acts by WAPA. In the absence of any formal consideration for other resources, and as hydroelectric peaking power became more valuable, the emphasis of the dam's operation became more firmly entrenched in the production of power.

One result of this emphasis is the day-to-day, constantly changing fluctuation in the river level. WAPA's determination to market hydropower at times of peak demand and higher revenues is the primary force controlling river flow. The increasing importance of power revenues was paralleled by a corresponding decrease in environmental and recreational concerns by both the Bureau of Reclamation and WAPA through the late 1980s, a situation bound to result in legal conflicts.

In 1989 a federal court finally recognized that WAPA's hydropower-generation policies and marketing criteria were degrading the Colorado River through Grand Canyon.[11] A landmark decision by the federal district court in Salt Lake City ruled that, before implementing new hydroelectric power contracts, WAPA had to prepare a full environmental impact statement concerning the effects of its power-marketing criteria on the Colorado River. This ruling was a significant victory for the environment and long-term conservation goals, because it initiated the process of bringing the operation of Glen Canyon Dam into compliance with the National Environmental Policy Act. For the first time since the dam was constructed, the Bureau of Reclamation and WAPA were being legally reminded of their mandate to consider natural and recreational resources in the formulation of their annual plans.

Figure 8.2. The daily fluctuations in water released from the dam are a product of hydropower marketing strategies used by the Western Area Power Administration. These fluctuations result in the creation of a bare "intertidal zone," or beach, between the average daily high and low flows, as seen in this photograph from the mid-1970s of the small river channel running left of the island at 209-Mile Rapid. *Steven W. Carothers*

The Money Machine

Deep inside Glen Canyon Dam, on a small deck overlooking the eight massive turbines generating hydroelectricity, is a small, unimposing machine. When the turbines are running, its lighted dials are constantly registering higher and higher numbers — the numbers of dollars earned by the sale of the dam's electric power. This is the money machine. A powerful symbol of but one of the three primary functions served by the dam, the money machine graphically establishes the magnitude of power revenues generated by the harnessing of the Colorado River.

More than four billion kilowatt-hours of electricity, an amount serving the needs of more than two million people, is produced by the dam each year, resulting in power revenues in excess of eighty million dollars annually. Glen Canyon Dam is the major component of the CRSP marketing and delivery network, generating approximately 10 percent of WAPA's total

kilowatt-hours and more than 70 percent of all federal hydropower produced in the intermountain West. It is no wonder, then, that WAPA pays such careful attention to the hour-by-hour management of the dam.

Glen Canyon Dam is only one feature of the overall CRSP, a wide-ranging federal public works program that has constructed hydroelectric dams, created massive irrigation projects, and provided for salinity-control measures throughout the river basin. Provisions of the Colorado River legislation of 1956 and 1968 stipulate that hydropower revenues from Glen Canyon Dam will be used not only to pay for running the dam, *but to help pay for the other participating projects of the CRSP.*

In essence, Glen Canyon Dam is the primary money machine for the entire CRSP. The dam must continually market hydropower to pay for upstream diversions and irrigation projects, many of which will never earn a cent of net revenue and represent, at best, poorly conceived government subsidies. The cumulative cost of these nonpower aspects of the project, as they are called, amounts to nearly half the total cost of the CRSP.

Figure 8.3. The money machine at Glen Canyon Dam constantly registers the enormous amounts of money earned by the sale of hydro-electric power generated at the dam. *Dugald Bremner*

Glen Canyon Dam has paid for itself several times over. Yet the dam must still maintain a large hydropower income because of the repayment schedule it is tied to, which in 1988 amounted to almost a billion dollars' worth of other CRSP financial responsibilities. The dam's legislated obligation to meet the repayment schedule has been and will continue to be the reason why power-generation revenues receive a higher priority than environmental and recreational concerns in the overall management framework at Glen Canyon Dam.

Figure 8.4. Glen Canyon Dam accounts for more than three-fourths of the hydropower revenues generated by the Colorado River Storage Project (CRSP) each year (*C*). Yet the costs of all power investments (*A, bottom*), the nonpower costs (*A, top*), and the annual operating costs of all power investments (*B*) which the revenues are legislatively mandated to repay dictate that Glen Canyon Dam will continue paying for the entire CRSP until well into the twenty-first century (*D* plus *E* equals *A*). As of 1988 only 40 percent of the total power investment had been repaid. Nonpower costs include irrigation projects and other activities. (Information courtesy Western Area Power Administration)

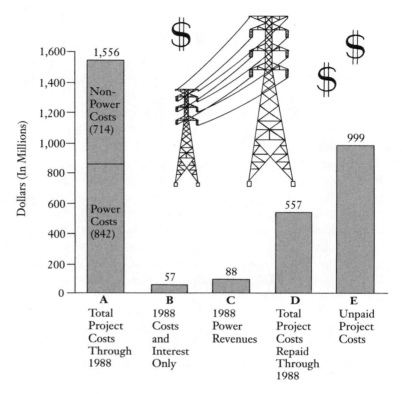

The Environmental Laws

The Colorado River Storage Project Act of 1956 and the Colorado River Basin Act of 1968 were enacted before the major environmental laws regulating the impacts of federal projects on public lands came into existence. The Endangered Species Act, the National Environmental Policy Act, and other environmental laws are just beginning to have an influence over how Glen Canyon Dam and the other dams in the basin are operated.

The Endangered Species Act of 1973 is one of the most powerful pieces of environmental legislation that could be brought to bear on the operation of Glen Canyon Dam. The act defines an endangered species as "one that is in danger of extinction throughout all or a significant portion of its range" and directs all federal agencies to ensure that their actions do not jeopardize the continued existence of any species designated as endangered by the U.S. Fish and Wildlife Service. Most importantly, the act stipulates that the actions of federal agencies may not result in the destruction or modification of habitat that is critical to the survival of an endangered species. Only the secretary of the interior, after consultation with the states involved, may designate an area as critical habitat — that area of land, water, and airspace required for the well-being and survival of a species.

The humpback chub was one of the first species classified as endangered when the act was first passed in 1973. The implications of its inclusion could eventually have major repercussions for the operation of Glen Canyon Dam because of the population of humpback chubs existing at the mouth of the Little Colorado River. As one of the endangered species directly influenced by the operation of the dam, the humpback chub and its ecological needs are in a powerful position to dictate how the dam may be operated in the future. The dilemma of how to manage the dam without jeopardizing the humpback chub has been the subject of intense scientific research and public controversy since the 1970s.

Bald eagles and peregrine falcons, as endangered species also affected by dam operation, may also play a role in determining how the dam is to be operated. The initial effect of the dam on these two endangered birds, though accidental, was apparently positive. Thus the dam has the potential to operate for both hydropower revenues and resource values.

Hand in hand with the Endangered Species Act is another important piece of congressional environmental legislation: the National Environmental Policy Act (NEPA) of 1969. The primary object of NEPA is to ensure that federal agencies consider the effects of their actions on the entire environment — biological, sociological, cultural, and economic — and eliminate, reduce, or, when appropriate, mitigate any negative impact. The key point of NEPA compliance is that it requires an assessment of the effects of major federal actions that are not strictly maintenance programs.[12] And part of this environmental assessment is to consider the influence of the proposed action on all organisms in the area to be affected, especially endangered species.

Although Glen Canyon Dam came into existence years before NEPA became law, major changes in the operation of the dam now fall under the jurisdiction of NEPA. The uprate and rewind of the generators at the dam from a capacity of 31,500 cfs to 33,200 cfs were subjected to a scaled-down version of the environmental assessment process in 1982 (an Environmental Assessment), with a finding of "no significant impact." The implications of this finding were still being debated when the

Figure 8.5. Environmental laws are only one of the factors determining the operating criteria for the dam. The number of agencies directly and indirectly involved in controlling dam operations has greatly increased since 1963, resulting in a complex hierarchy of priorities. Even small changes in the operating criteria can have profound effects on the living and nonliving systems of the downstream river.

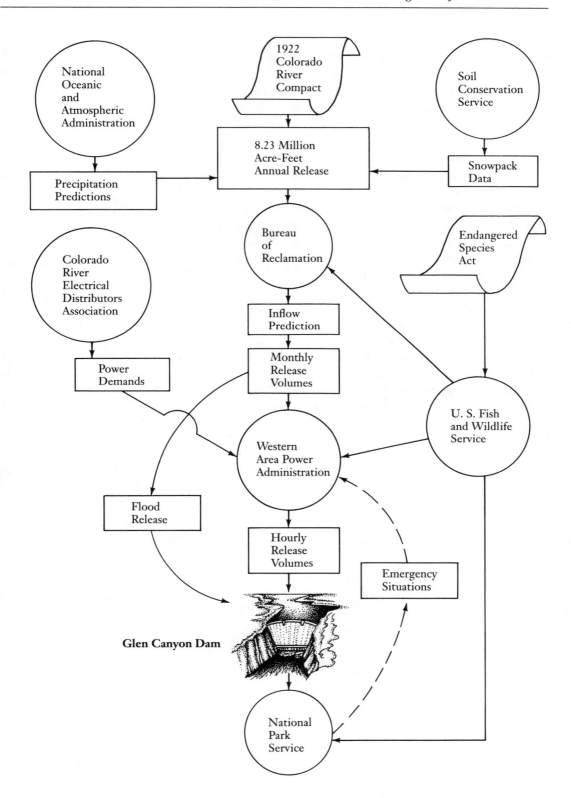

Glen Canyon Dam

secretary of the interior called for the initiation of a full
Environmental Impact Statement (EIS) on the operations of
Glen Canyon Dam in 1989. To provide sufficient data for the
EIS, the secretary further directed an expansion of the Glen
Canyon Environmental Studies research program in 1990. This
process may affect the operation of the dam through the twenty-
first century and could eventually help push it in a more
environmentally conscious direction.

There is no doubt that a new day is dawning for the Colorado
River in Grand Canyon. Progress is slowly being made on the
environmental front, as seen in the late 1980s by federal district
court decisions and the initiation of the EIS process. And this
progress is occurring in spite of the competing mandates and
priorities of the varied agencies managing the operation of Glen
Canyon Dam and the unfortunate delay in initiating the
provisions of important environmental legislation. But more
work is needed: the reevaluation of the operating criteria of the
dam, the development of a new focus on management goals as
they concern the downstream ecosystem, and the integration of
agency priorities and mandates into an overall, cohesive set of
environmentally sensitive objectives for the Colorado River of
the future. The future management plan for the river will no
doubt provide for hydropower revenues while protecting the
natural values of Grand Canyon National Park. These goals are
not mutually exclusive.

EPILOGUE
The River of the Future

In the summer of 1858 a river expedition of an unusual sort was attempting to travel into the Grand Canyon. The War Department had authorized Lieutenant Joseph C. Ives to explore the lower Colorado River and determine how far upstream it was navigable for steamboats. Had the effort succeeded, it would have been the first recorded river expedition to venture into the Grand Canyon, in spite of Ives's unorthodox approach of traveling upstream by boat.

The steamboat *U.S. Explorer*, a sternwheeler, was used by Ives to travel up the lower Colorado River. Two months of backbreaking labor fighting rapids, shifting sandbars, and swift currents brought the expedition to Black Canyon, some 350 miles upstream from the river's mouth on the Gulf of California. There the steamboat struck a submerged rock near the present site of Hoover Dam and was wrecked beyond repair. Lieutenant Ives wisely concluded that steamboats would not be able to proceed upriver above Black Canyon.[1]

Expedition members continued eastward on foot, proceeding up the river and across the plateaus into the Grand Canyon region. Lieutenant Ives eventually visited the Colorado River at the mouth of Diamond Creek and reached the Havasupai villages in Cataract Canyon before turning back. Either the expedition did not go well, or the participants were not impressed with what they saw, for one later described Grand Canyon as "the gate of Hell." Lieutenant Ives summed up his view of the Colorado River through Grand Canyon in what is now a classic statement:

> The region is, of course, altogether valueless. It can be approached only from the south, and after entering it there is nothing to do but leave. Ours has been the first, and will doubtless be the last, party of whites to visit this profitless locality. It seems intended by nature that the Colorado River, along the greater portion of its lonely and majestic way, shall be forever unvisited and undisturbed.[2]

If Lieutenant Ives could see the Colorado River of today, he would doubtless be astounded to learn how everything had changed. Large dams at either end of the Grand Canyon control the Colorado's flow, and thousands of people travel down the river each year to view its "lonely and majestic way." Even the plants and animals found in and along the river have changed.

Forever Undisturbed: The Hollow Legacy

Change in the natural world of the Colorado River through Grand Canyon is nothing new, as events from the ice age to the present have clearly demonstrated. Yet most of the changes to the Colorado River ecosystem since Lieutenant Ives first saw it have been caused by man. These changes have resulted in a river that has a different look, a different feel, a different flora and fauna, and a different pace of life.

The natural process of change along the river has been greatly accelerated by man's actions. Natural cycles of replenishment and renewal have been interrupted and in many cases thrown wildly out of balance. An entirely new river has recently emerged, one with a radically different sediment budget, water cycle, and food web. The river of the present and the future is no longer natural, but instead is *naturalized*, a blend of the old and the new, a mixture of native and introduced organisms and natural and artificial processes.

That is not to imply that the new, naturalized river is of any less or any more value. Aesthetically, biologically, and spiritually, the Colorado River through Grand Canyon is a world-class treasure that more than deserves the attention and praise it receives. The international prestige of the Grand Canyon and its river demands that it be managed in a way that balances our use of the river with its unparalleled values, perhaps even in a way that enhances the new naturalized values of the river corridor.

The political tug-of-war among the various agencies, laws, and mandates that determines the environmental health of the river corridor is only part of the problem. The greatest management difficulty results from the lack of a clear vision as to what the new river corridor should be managed for. The National Park Service is in the position to determine management goals and establish resource priorities but has not done so.

The policy of the National Park Service is to "preserve and manage the Parks in their natural state for the enjoyment of future generations" as vignettes of primitive America.[3] This approach may be wholly inappropriate for the complex problems confronting the management of the Colorado River in Grand Canyon, for along with present National Park Service policies, this policy provides no guidance for coping with naturalized ecosystems. Indeed, naturalized ecosystems are not yet officially recognized by the National Park Service. Management goals stress the removal of non-native organisms and the nurturing, or even reintroduction, of natural ones. There are no provisions in today's natural resource management policy for accepting, managing, and protecting exotic species, even when these species are permanently established and beneficial to native species.

In 1961 Secretary of the Interior Stewart Udall commissioned a document to provide the National Park Service with guidance in the management of wildlife and natural processes in the parks. This document, known as the Leopold Report after commission chairman A. Starker Leopold, drove home the message that exotic species and unnatural processes did not belong in the parks. Park management, it reasoned, should be limited to native species and not exotic ones.

The Leopold Report was a milestone of natural resource management in the parks, setting the tone for resource decisions made throughout the remainder of the twentieth century. Although the report was a godsend for the enlightened management of the national parks, the majority of which exhibited relatively unaltered ecosystems, it did not recognize as exceptions those park ecosystems that had changed or would soon change beyond the point of no return.

The Leopold Report was submitted in 1963, ironically the same year that the gates of Glen Canyon Dam were first closed and the Colorado River in Grand Canyon ceased to be a naturally regulated park ecosystem. Although the National Park Service had introduced many non-native species into the park before the report labeled exotics as inappropriate, it was not until Glen Canyon Dam changed the river environment that exotics began to assume dominance. When this trend first became apparent, the National Park Service responded by ignoring the dilemma. As time went on, the service never established a set of coherent natural resource management goals for the river, precisely because of this indecision about the overwhelming dominance of introduced species. Even the 1989 version of the Leopold Report stressed the maintenance or restoration of native biota. While admitting that in some instances the re-creation of natural ecosystems was impossible, the updated report offered little guidance in the management of naturalized environments within the national park system.[4]

A new vision is needed, both inside and outside the National Park Service, for ecosystem management along the Colorado River in Grand Canyon. It is particularly important that the National Park Service, as the highly visible flagship agency shepherding the Grand Canyon, espouse this new vision. If Glen Canyon Dam is here to stay — it seems beyond reason that the dam and the Colorado River Storage Project would vanish overnight — then a new perspective on river management must be developed. Considerations influencing this new perspective should include environmental concerns, water-delivery requirements, hydropower revenues, and recreation.

Restoring the river to prehistoric conditions, the de facto goal of the National Park Service, is simply not an option. An expensive re-creation of the river of yesterday could be fabricated by drawing warmer water from the surface of Lake Powell, reinjecting sediment into the river below the dam, and tailoring water releases to mimic natural cycles of flood and drought. But even with all of this effort, the river could not be truly restored. The extirpated species of native fish would probably not return, the Colorado River Compact would not go away, and water and power needs would continue to increase. Political and economic realities demand a new definition of the river of the future. The agencies and interests that control Glen Canyon Dam must work together to define the new river — within a National Park framework — and to maintain the integrity of the naturalized river ecosystem.

Natural or naturalized, the values of the new river are important enough not to be neglected by mismanagement through indecision. Most of the proper tools are at hand, save a single, unifying legislative act, to accomplish this task.

Possibilities for the Future

Because the future of the Grand Canyon section of the river is intimately tied to the future of the entire river system, from its headwaters to the Gulf of California, a broad perspective is in order. The river of the future may be defined, in part, by a basin-wide Environmental Impact Statement that is needed to address the cumulative, integrated effects of all water-management practices on the river system. This basin-wide analysis would almost certainly need to be initiated and authorized by Congress.

The river of the future might be guided by a "Colorado River Environmental Conservation Act," a much-needed congressional mandate directing a holistic approach to river and dam management throughout the entire Colorado River drainage. This much-needed act would firmly establish and articulate the different goals and priorities for the management of the different reaches of the river. Likewise, it would define the operating criteria of the dams and reservoirs. The different agencies involved would be legislatively required to cooperate in the accomplishment of these goals.

This future law would specifically direct the secretary of the interior to coordinate the operation of Glen Canyon Dam, and indeed all dams in the Colorado River basin, so that downstream resources would be protected, mitigated, and enhanced. The act would redefine, in no uncertain terms, the intent of the Colorado River Storage Act of 1956 and the Colorado River Basin Project Act of 1968 that environmental and recreational criteria would have a higher priority than power generation throughout the system. And finally, the act would reassess the debt-repayment schedule outlined by the acts, in order to create a more flexible schedule that would allow for environmental protection and enhancement. No revision of the conditions of the Colorado River Compact of 1922 is envisioned or needed at this time, but the dreaded possibility exists that even these "Ten Commandments" may need to be reevaluated in the twenty-first century.

Rather than dictating the exact, day-to-day operating criteria for each dam, the act would create a Colorado River Environmental Council to oversee that responsibility. The council would meet on a regular basis to examine the changing nature of the river system and the resulting resource needs. The council would be empowered to determine the means to minimize environmental impacts and maximize resource values, as well as to enforce these decisions with the different agencies involved. It would be the obligation of the council continually to redefine the goals and objectives of river and dam management, in a program that would be responsive to unforeseen changes and needs.

At the heart of this program would be the concept of adaptive management. Rather than being locked into a rigid hierarchy of management and operating criteria to protect hydropower revenues or natural resources, resources that are constantly shifting and in need of a fresh perspective, adaptive management takes the flexible approach. Long-term power contracts, for example, would be inappropriate and unnecessary for the Glen Canyon Dam of the future.

This act of the future would call for the use of a certain percentage of the power revenues obtained from the sale of hydroelectric power to finance the activities of the council. These revenues would also be used to support an organized program of long-term research and monitoring on economic, environmental, and recreational resource questions, since management must be guided by up-to-date information. Public involvement would be identified as a key element of the information feedback system helping to direct the activities of the council.

The earmarking of revenues for environmental considerations, operational flexibility, and adaptive management will exact a price. It will certainly involve a slightly higher cost for hydroelectric power, which is inevitable if the natural values of the Colorado River corridor receive their rightful priority. The price of hydroelectric power from Glen Canyon Dam is so highly subsidized — power from the dam is among the cheapest in the United States — that slightly higher power costs are not unreasonable.

Despite the unavoidable cost, four criteria governing the flows released from Glen Canyon Dam would be mandated by this hypothetical law of the future. These criteria would all be directed at eliminating or reducing the two most damaging effects of the dam on the downstream environment — rapid fluctuations in water level and uncontrolled flooding — while fulfilling the requirement to release 8.23 maf each year. First, minimum flows should be identified, based on sound scientific studies, which would be most beneficial to the downstream environment. Second, maximum flows should be defined according to the same criteria, with the ultimate goal of greatly reducing the overall magnitude of fluctuations. Third, the rate of fluctuations (the time it takes to go from minimum to maximum release) should be dampened, allowing for a more gentle transition from low to high. And finally, alternative plans should be developed to cope with emergency power demands or unexpected surplus water that could result in unplanned downstream flooding. In all cases, unplanned flooding must be avoided.

As is sometimes the case, a compelling and successful precedent exists for congressional passage of just such an act. In

1980 Congress passed the Pacific Northwest Power Planning and Conservation Act, which was nothing less than a revolutionary piece of legislation for river management. The Northwest Power Planning Act, as it is called, is an interstate compact among the states of Washington, Oregon, Idaho, and Montana, as well as Indian tribes and other organizations. Representatives from each of the states form the Pacific Northwest Electric Power and Conservation Planning Council,[5] a group empowered to produce a long-term plan for river and dam management as well as to reevaluate continually the changing needs of the river's resources. The council, with an operating budget of approximately seven million dollars a year, is funded by a percentage of hydroelectric power revenues collected by the Bonneville Power Administration (similar to the Western Area Power Administration), which markets power in the Pacific Northwest. The council produces an annual water budget for the river system, supports an intensive program of environmental research and monitoring, and incorporates public input into every facet of its operation.

In essence, the act called for the development of an integrated, long-term plan for hydroelectric management and environmental protection and restoration of the Columbia River basin which will provide both inexpensive electricity and restore the natural values of the river ecosystem. Widespread ecological disruption of the Columbia River was the result of years of poorly planned hydroelectric development. The act specified that this disruption must now be balanced, and it called for the mitigation, replacement, and enhancement of natural habitats and processes, as well as for the restoration of species extirpated from parts of the river system.

Once the full recommendations of the Northwest Power Planning Council are enacted, the Columbia River will become the target of one of the largest environmental restoration efforts in the world. The full cost of this restoration could amount to more than fifty million dollars annually, paid for by the consumers of electricity. The Colorado River deserves no less.

Although the Colorado River in Grand Canyon may seem timeless compared to the small scale of our lifetimes, the river is anything but unchanging. Both natural and man-made changes will continue to occur in the river corridor. Man has taken the upper hand in determining the future of the river, but this influence need not be entirely negative. Without seriously reducing hydropower revenues, the dam can easily be operated to meet the two most important ecological needs of the new river: stabilizing the aquatic and riparian ecosystems and striving for a sediment balance that is at or near equilibrium.

If we must assume that the dams and reservoirs are here to stay, their influence should be universally beneficial. We now possess the capability to maintain, restore, and even enhance the river of the future via the enlightened release of water from Glen Canyon Dam. Such a goal is not unreasonable for one of the world's greatest natural treasures. The profound wonder of the Colorado River through Grand Canyon, with its ability to humble, heal, and illuminate the human experience through a vibrant encounter with the natural world, may be even more important to future generations than it is today.

APPENDIX A
Plants and Animals of the River Corridor

Table A.1   Primary Woody Plant Species of the River Corridor

Habitat Zone and Common Name	Scientific Name	Abundance
SEEPS AND SPRINGS		
Native plants		
Maidenhair fern	*Adiantum capillus-veneris*	Common
Crimson monkeyflower	*Mimulus cardinalis*	Common
Giant helleborine orchid	*Epipactis gigantea*	Uncommon
Golden columbine	*Aquilegia chrysantha*	Common
Bear grass	*Nolina* sp.	Common
Sawgrass	*Cladium californicum*	Common
Birchleaf buckthorn	*Rhamnus betulaefolia*	Uncommon
Fremont cottonwood	*Populus fremontii*	Common
Velvet ash	*Fraxinus pennsylvania*	Common
Single-leaf ash	*Fraxinus annomala*	Common
Introduced plants	None	
OLD HIGH-WATER ZONE		
Native plants		
Honey mesquite	*Prosopis glandulosa*	Abundant; has invaded new high-water zone
Catclaw acacia	*Acacia greggii*	Common; has invaded new high-water zone
Apache plume	*Fallugia paradoxa*	Common in Marble Canyon
Netleaf hackberry	*Celtis reticulata*	Uncommon
Redbud	*Cercis occidentalis*	Common in Marble Canyon
Desert isocoma	*Haplopappus acradenius*	Common
Scrub oak	*Quercus turbinella*	Common in Glen Canyon
Rabbitbrush	*Chrysothamnus* spp.	Common in Glen Canyon
Introduced plants	None	

Table A.1   (*Continued*)

Habitat Zone and Common Name	Scientific Name	Abundance
NEW HIGH-WATER ZONE		
Native plants		
Coyote willow	*Salix exigua*	Abundant
Goodding willow	*Salix gooddingii*	Uncommon
Arrowweed	*Tessaria sericea*	Abundant
Desert broom	*Baccharis sarathroides*	Common below Lava Falls
Waterweed	*Baccharis sergiloides*	Uncommon
Seep willow	*Baccharis glutinosa*	Common
Emory seep willow	*Baccharis emoryi*	Uncommon
Longleaf brickellia	*Brickellia longifolia*	Common
Reed, or carrizo	*Phragmites australis*	Common
Cattail	*Typha* spp.	Common
Spiny aster	*Aster spinosus*	Uncommon
Cottonwood	*Populus fremontii*	Rare and localized
Introduced plants		
Tamarisk, or saltcedar	*Tamarix chinensis*	Abundant
Camelthorn	*Alhagi camelorum*	Common
Russian olive	*Elaeagnus angustifolia*	Rare
Elm	*Ulmus minor*	Rare in Glen Canyon

Desert species that have invaded the riparian zones or are adjacent to them

Native plants		
Creosote bush	*Larrea tridentata*	Abundant below Havasu Creek
Brittle bush	*Encelia farinosa*	Abundant below Marble Canyon
Ocotillo	*Fouquieria splendens*	Abundant below Havasu Creek
Pepper grass	*Lepidium montanum*	Abundant in Marble Canyon
Barrel cactus	*Ferocactus acanthodes*	Abundant below River Mile 26
Prickly pear cactus	*Opuntia* spp.	Abundant
Cholla cactus	*Opuntia* spp.	Common below Lava Falls?
Introduced plants	None	

SUGGESTED READINGS

Phillips, B. G., A. M. Phillips, III, and M.A.S. Bernzott. *Annotated Checklist of Vascular Plants of Grand Canyon National Park.* Monograph No. 7. Grand Canyon, Ariz.: Grand Canyon Natural History Association, 1987.

Turner, R. M., and M. M. Karpiscak. *Recent Vegetation Changes Along the Colorado River Between Glen Canyon Dam and Lake Mead, Arizona.* U.S. Geological Survey Professional Paper 1132. Washington, D.C.: U.S. Government Printing Office, 1980.

Table A.2 Amphibians and Reptiles of the River Corridor

Common Name *Scientific Name*	*Abundance*	*Habitat*[a]
AMPHIBIANS		
Rocky Mountain toad		
Bufo woodhousei	Abundant	Riparian
Red-spotted toad		
Bufo punctatus	Abundant	Desert, riparian
Canyon treefrog		
Hyla arenicolor	Common	Tributaries
Leopard frog		
Rana pipiens	Rare?	Riparian
LIZARDS		
Desert banded gecko		
Coleonyx variegatus	Rare	Riparian
Chuckwalla		
Sauromalus obesus	Common	Cliff, desert, riparian
Zebra-tailed lizard		
Callisaurus draconoides	Rare	Desert, riparian
Black-collared lizard		
Crotaphytus bicinctores	Uncommon	Desert, riparian
Yellow-backed desert spiny lizard		
Sceloporus magister	Abundant	Cliff, desert, riparian
Side-blotched lizard		
Uta stansburiana	Abundant	Desert, riparian
Tree lizard		
Urosaurus ornatus	Abundant	Cliff, riparian
Desert horned lizard		
Phrynosoma platyrhinos	Rare	Desert, riparian
Northern whiptail		
Cnemidophorous tigris	Abundant	Desert, riparian
Gila monster		
Heloderma suspectum	Rare	Desert, riparian

Table A.2 (*Continued*)

Common Name *Scientific Name*	*Abundance*	*Habitat*[a]
SNAKES		
Western blind snake *Leptotyphlops humilis*	Rare	Desert, riparian
Ringneck snake *Diadophis punctatus*	Rare	Desert, riparian
Red racer, or coachwhip *Masticophis flagellum*	Uncommon	Desert, riparian
Desert striped whipsnake *Masticophis taeniatus*	Uncommon	Desert, riparian
Mohave patch-nosed snake *Salvadora hexlepis*	Uncommon	Desert, riparian
Sonoran gopher snake *Pituophis melanoleucus*	Uncommon	Desert, riparian
California kingsnake *Lampropeltus getulus*	Common	Desert, riparian
Long-nosed snake *Rhinocheilus lecontei*	Rare	Desert, riparian
Sonoran lyre snake *Trimorphodon biscutatus*	Rare	Desert, riparian
Desert night snake *Hypsiglena torquata*	Rare	Desert, riparian
Western diamondback rattlesnake *Crotalus atrox*	Rare	Desert, riparian
Speckled rattlesnake *Crotalus mitchelli*	Rare	Desert, riparian
Black-tailed rattlesnake *Crotalus molossus*	Rare	Desert, riparian
Grand Canyon rattlesnake *Crotalus viridis abyssus*	Common	Desert, Riparian

Table A.2 (*Continued*)

*a*Cliff = vertical rock faces, broken rocky outcrops; Desert = desertscrub habitats away from the riparian zone of the river; Riparian = tamarisk, mesquite, sandy, or shoreline habitats; Tributaries = confluence of clear, flowing side streams

SUGGESTED READINGS

Miller, D. A., R. A. Young, T. W. Gatlin, and J. A. Richardson. *Amphibians and Reptiles of the Grand Canyon National Park.* Monograph No. 4. Grand Canyon, Ariz.: Grand Canyon Natural History Association, 1982.

Stevens, L. *The Colorado River in Grand Canyon: A Guide.* Flagstaff, Ariz.: Red Lake Books, 1983.

Warren, P. L., and C. R. Schwalbe. *Lizards Along the Colorado River in Grand Canyon National Park: Possible Effects of Fluctuating River Flows.* Glen Canyon Environmental Studies Technical Report. Salt Lake City: Bureau of Reclamation, 1986.

Table A.3   Nesting Birds of the River Corridor

Common Name Scientific Name	Nesting Abundance[a]	Nesting Habitat[b]
Black-crowned night-heron *Nycticorax nycticorax*	Uncommon	NHWZ
Mallard *Anas platyrhynchos*	Uncommon	NHWZ
Gambel's quail *Callipepla gambelii*	Uncommon	OHWZ, desert
American coot *Fulica americana*	Rare	NHWZ
Spotted sandpiper *Actitis macularia*	Common	Open sand
Mourning dove *Zenaida macroura*	Common	NHWZ, OHWZ
Greater roadrunner *Geococcyx californianus*	Uncommon	Desert
Western screech-owl *Otus kennicottii*	Rare	NHWZ, OHWZ
Great horned owl *Bubo virginianus*	Uncommon	Cliff, desert
White-throated swift *Aeronautes saxatalis*	Common	Cliff
Black-chinned hummingbird *Archilochus alexandri*	Common	NHWZ

Table A.3   (*Continued*)

Common Name Scientific Name	Nesting Abundance[a]	Nesting Habitat[b]
Costa's hummingbird *Calypte costae*	Common	OHWZ, desert
Ladder-backed woodpecker *Picoides scalaris*	Rare	OHWZ, desert
Willow flycatcher *Empidonax traillii*	Rare	NHWZ
Black phoebe *Sayornis nigricans*	Common	Cliff
Say's phoebe *Sayornis saya*	Fairly common	Cliff
Ash-throated flycatcher *Myiarchus cinerascens*	Fairly common	NHWZ, OHWZ
Western kingbird *Tyrannus verticalis*	Uncommon	NHWZ, OHWZ
Violet-green swallow *Tachycineta thalassina*	Common	Cliff
Cliff swallow *Hirundo pyrrhonota*	Extirpated	Cliff
Cactus wren *Campylorhynchus* *brunneicapillus*	Uncommon	Desert
Rock wren *Salpinctes obsoletus*	Common	Desert
Canyon wren *Catherpes mexicanus*	Common	Cliff, desert
Bewick's wren *Thryomanes bewickii*	Common	NHWZ, OHWZ
Marsh wren *Cistothorus palustris*	Rare	NHWZ
American dipper *Cinclus mexicanus*	Fairly common	Tributaries
Blue-gray gnatcatcher *Polioptila caerulea*	Common	NHWZ, OHWZ
Black-tailed gnatcatcher *Polioptila melanura*	Rare	Desert
Northern mockingbird *Mimus polyglottos*	Uncommon	OHWZ, desert
Phainopepla *Phainopepla nitens*	Uncommon	OHWZ
European starling[c] *Sturnus vulgaris*	Uncommon	Urban
Bell's vireo *Vireo bellii*	Common	NHWZ, OHWZ

Table A.3   (*Continued*)

Common Name Scientific Name	Nesting Abundance[a]	Nesting Habitat[b]
Lucy's warbler *Vermivora luciae*	Common	NHWZ, OHWZ
Yellow warbler *Dendroica petechia*	Fairly common	NHWZ
Common yellowthroat *Geothlypis trichas*	Fairly common	NHWZ
Yellow-breasted chat *Icteria virens*	Common	NHWZ, OHWZ
Summer tanager *Piranga rubra*	Rare	NHWZ
Black-headed grosbeak *Pheucticus melanocephalus*	Rare	NHWZ
Blue grosbeak *Guiraca caerulea*	Fairly common	NHWZ, OHWZ
Lazuli bunting *Passerina amoena*	Uncommon	NHWZ, OHWZ
Indigo bunting *Passerina cyanea*	Rare	NHWZ, OHWZ
Rufous-crowned sparrow *Aimophila ruficeps*	Rare	Desert
Black-throated sparrow *Amphispiza bilineata*	Fairly common	Desert
Red-winged blackbird *Agelaius phoeniceus*	Fairly common	NHWZ
Great-tailed grackle *Quiscalus mexicanus*	Fairly common	NHWZ
Brown-headed cowbird *Molothrus ater*	Common	NHWZ, OHWZ
Hooded oriole *Icterus cucullatus*	Fairly common	NHWZ, OHWZ
Northern oriole *Icterus galbula*	Common	NHWZ
House finch *Carpodacus mexicanus*	Common	NHWZ, OHWZ
Lesser goldfinch *Carduelis psaltria*	Fairly common	NHWZ
House sparrow[c] *Passer domesticus*	Uncommon	Urban

Note: This table primarily includes species nesting in riparian vegetation; also listed are birds nesting in desertscrub or cliff habitats immediately adjacent to the river corridor, as those species can be seen and heard from a boat in midstream. Some cliff-nesting species,

Table A.3   *(Continued)*

however, such as red-tailed hawk (*Buteo jamaicensis*), peregrine falcon (*Falco peregrinus*), and American kestrel (*Falco sparverius*), have not been listed here. These three birds are frequently seen along the river but usually nest on high cliffs that are outside of what is usually thought of as the river corridor.

[a]Summer abundance during the nesting season: Common = easily found and always to be seen; Fairly common = found in small numbers in the right habitat; Uncommon = may or may not be found easily in small numbers in the right habitat; Rare = unpredictable, not always to be found even in the right habitat; Extirpated = historically nested, but no longer occurs as a nesting species.

[b]NHWZ = new high-water zone, dominated by tamarisk; OHWZ = old high-water zone, dominated by honey mesquite; Cliff = Vertical rockfaces immediately adjacent to river; Desert = desertscrub habitats, including talus, immediately adjacent to riparian zone; Open sand = sandy habitat largely free of vegetation; Tributaries = confluence of clear, perennial side streams; Urban = near human habitations, Lees Ferry, Phantom Ranch
[c]Introduced

SUGGESTED READINGS

Brown, B. T., S. W. Carothers, and R. R. Johnson. *Grand Canyon Birds: Historical Notes, Natural History, and Ecology.* Tucson: University of Arizona Press, 1987.

Brown, B. T., and R. R. Johnson. *Fluctuating Flows from Glen Canyon Dam and Their Effect on Breeding Birds of the Colorado River.* Glen Canyon Environmental Studies Technical Report. Salt Lake City: Bureau of Reclamation, 1987.

Table A.4   Mammals of the River Corridor

Common Name Scientific Name	Abundance	Habitat
BATS		
Yuma myotis *Myotis yumanensis*	Common	Desert, riparian
Small-footed myotis *Myotis californicus*	Abundant	Desert, riparian
Silver-haired bat *Lasionycteris noctivagans*	Rare	Desert, riparian
Western pipistrelle *Pipistrellus hesperus*	Abundant	Desert, riparian
Big brown bat *Eptesicus fuscus*	Uncommon	Desert, riparian

Table A.4 *(Continued)*

Common Name Scientific Name	Abundance	Habitat
Townsend's big-eared bat *Plecotus townsendii*	Rare	Desert, riparian
Pallid bat *Antrozous pallidus*	Common	Desert, riparian
American free-tailed bat *Tadarida brasiliensis*	Uncommon	Desert, riparian
RODENTS		
Rock squirrel *Spermophilus variegatus*	Common	Desert, riparian
Harris' antelope squirrel *Ammospermophilus harrisii*	Rare	Desert, riparian
White-tailed antelope squirrel *Ammospermophilus leucurus*	Uncommon	Desert, riparian
Merriam's kangaroo rat *Dipodomys merriami*	Uncommon	Desert, riparian
Cliff chipmunk *Eutamias dorsalis*	Uncommon	Desert, riparian
Long-tailed pocket mouse *Perognathus formosus*	Uncommon	Desert, riparian
Rock pocket mouse *Perognathus intermedius*	Common	Desert, riparian
Beaver *Castor canadensis*	Common	Aquatic, riparian
Western harvest mouse *Reithrodontomys megalotis*	Uncommon	Desert, riparian
Canyon mouse *Peromyscus crinitus*	Abundant	Desert, riparian
Cactus mouse *Peromyscus eremicus*	Abundant	Desert, riparian
Deer mouse *Peromyscus maniculatus*	Uncommon	Riparian
Brush mouse *Peromyscus boylii*	Uncommon	Desert, riparian
Pinyon mouse *Peromyscus truei*	Rare	Desert, riparian
White-throated woodrat *Neotoma albigula*	Abundant	Desert
Desert woodrat *Neotoma lepida*	Abundant	Desert
Muskrat *Ondatra zibethicus*	Rare	Aquatic, riparian

Table A.4   (*Continued*)

Common Name Scientific Name	Abundance	Habitat
CARNIVORES		
Coyote		
Canis latrans	Uncommon	Desert, riparian
Gray fox		
Urocyon cineroargentatus	Uncommon	Desert, riparian
Ringtail		
Bassariscus astutus	Common	Desert, riparian
Raccoon		
Procyon lotor	Rare	Riparian
Western spotted skunk		
Spilogale gracilis	Common	Desert, riparian
River otter		
Lutra canadensis	Rare?	Aquatic; extirpated?
Mountain lion		
Felis concolor	Rare	Desert, riparian
Bobcat		
Lynx rufus	Rare	Desert, riparian
UNGULATES		
Mule deer		
Odocoileus hemionus	Common	Desert, riparian
Bighorn sheep		
Ovis canadensis	Common	Desert, riparian
Feral burro		
Equus asinus	Rare	Desert, riparian; introduced species found only in upper Lake Mead area

Note: The river corridor includes the aquatic, riparian, and immediately adjacent desertscrub habitats. The larger mammals are highly mobile and range widely through both desert and riparian habitats.

SUGGESTED READINGS

Butterfield, K., B. T. Brown, R. R. Johnson, and N. Czaplewski. *A Checklist: Mammals of the Grand Canyon Region*. Grand Canyon, Ariz.: Grand Canyon Natural History Association, 1981.

Hoffmeister, D. F. *Mammals of Arizona*. Tucson: University of Arizona Press; Arizona Game and Fish Department, 1986.

Ruffner, G. A., N. J. Czaplewski, and S. W. Carothers. "Distribution and Natural History of Some Mammals from the Inner Gorge of the Grand Canyon, Arizona." *Journal of the Arizona-Nevada Academy of Science* 13 (1978): 85–91.

Stevens, L. *The Colorado River in Grand Canyon: A Guide*. Flagstaff, Ariz.: Red Lake Books, 1983.

Table A.5   Status and Distribution of Colorado River Fishes in Grand Canyon

Common Name Scientific Name	Status[a]	Distribution and Comments
NATIVE SPECIES (8)		
Roundtail chub *Gila robusta*	Extirpated	May never have been common
Bonytail chub *Gila elegans*[b]	Extirpated	May never have been common
Humpback chub *Gila cypha*[b]	Local	Small population persists at Little Colorado River confluence
Colorado squawfish *Ptychocheilus lucius*[b]	Extirpated	Apparently in population decline before 1963
Speckled dace *Rhinichthys osculus*	Common	Mostly in tributaries, some adults in main stream
Razorback sucker *Xyrauchen texanus*	Extirpated?	Adult female captured and released in 1984; possibly in decline before 1963
Flannelmouth sucker *Catostomus latipinnis*	Abundant	Widespread
Bluehead sucker *Catostomus discobolus*	Common	Widespread
INTRODUCED SPECIES (20)		
Utah chub *Gila atraria*	Accidental	One record from Lees Ferry, not established
Coho salmon *Oncorhynchus kisutch*	Accidental	Did not become established
Rainbow trout *Salmo gairdneri*	Abundant	Introduced in 1920; stocking continues at Lees Ferry
Cutthroat trout *Salmo clarki*	Rare	Introduced at Lees Ferry in 1978; once abundant, now fished out
Brown trout *Salmo trutta*	Fairly common	Local; at Lees Ferry, Bright Angel Creek, and a few other tributaries
Brook trout *Salvelinus fontinalis*	Common	Ubiquitous, most common in Lees Ferry area
Carp *Cyprinus carpio*	Uncommon to rare	Ubiquitous; present since late 1800s; once abundant after the dam, has since declined in numbers
Golden shiner *Notemigonus crysoleucus*	Accidental	Few records, not yet established in canyon
Virgin River spinedace *Lepidomeda mollispinis*	Accidental	One record from Paria River in 1972, probably accidentally introduced

Table A.5   (*Continued*)

Common Name Scientific Name	Status[a]	Distribution and Comments
Woundfin *Plagopterus argentissimus*	Accidental	Introduced into Paria River in 1972 but did not become established
Red shiner *Notropis lutrensis*	Accidental	Apparently introduced
Fathead minnow *Pimephales promelas*	Common	Locally found in many tributaries
Channel catfish *Ictalurus punctatus*	Common	Ubiquitous but locally most common in warm-water tributaries, Little Colorado River, Kanab Creek; introduced and well established by early 1900s
Black bullhead *Ictalurus melas*	Rare	Occasionally found in tributaries
Rio Grande killifish *Fundulus zerbrinus*	Common	Locally found in Unkar, Royal Arch, and Kanab creeks and a few others
Striped bass *Morone saxatilis*	Rare	Occasionally found in river but not established there; common in Lake Mead
Largemouth bass *Micropterus salmoides*	Rare	Occasionally found in river, but all have come upstream from lake Mead; not established
Green sunfish *Chaenobryttus cyanellus*	Rare	Sometimes found in lower canyon; not established
Bluegill sunfish *Lipomis macrochirus*	Rare	Sometimes found in lower canyon; not established
Walleye *Stizostedion vitreum*	Accidental	One record from Lees Ferry, where it was probably introduced from Lake Powell

[a]Abundant = easily captured and always present in large numbers; Common = easily captured although not always present in large numbers; Fairly common = occasionally captured, but not unexpected; Locally common = captured easily in specific areas and sometimes present in numbers; Uncommon = captured with difficulty in appropriate habitat, not present in numbers; Rare = captured with extreme difficulty in appropriate habitat, always present in small numbers; Accidental = one or two specimen records, isolated incidences of bait-bucket releases, relatively unsuccessful transplants, or individuals dispersing upstream from lake Mead.

[b]A legally designated and protected species in danger of extinction.

APPENDIX B
Chronology

Table B.1   Events in the Natural History of the Colorado River in Grand Canyon or Our Interpretation of It

Date	Event
1 million years ago	Colorado River has cut to its approximate present depth, at least in the western Grand Canyon.
140,000 years ago	The most recent lava flows near Toroweap dam and impound the river to a depth of perhaps 225 feet, creating a lake more than 40 miles long.
11,000 years ago	The "Little Ice Age" ends, causing the climate of the Southwest gradually to become warmer and drier.
1826	James Ohio Pattie and others trap for beaver along the Colorado River, possibly near the Grand Wash Cliffs.
1869–72	Major John Wesley Powell completes one exploratory trip down the river through Grand Canyon and part of another.
1881	The last California condor in Grand Canyon is shot by prospectors along the Colorado River near Pearce Ferry.
1884	An ungauged flood of approximately 300,000 cfs becomes the largest flood in historical times.
1890	The Stanton Expedition takes almost four hundred photographs of the river corridor, an invaluable record of the river's condition at that time.
1896	The Flavell-Montez Expedition traps furbearers on a river trip through the length of the canyon.
1896–97	Nathaniel Galloway traps furbearers on a river trip through the length of the canyon.
1908	Grand Canyon National Monument is created by President Theodore Roosevelt.
1911–12	Kolb brothers complete a successful downriver trip through the canyon, provide historical records of fish and wildlife, and make a motion picture of their experiences.
1919	Grand Canyon National Park is created by Congress.
1920s	Tamarisk invades the river corridor.
1921	Suspension (black) bridge built across the river at Phantom Ranch.

Table B.1 (*Continued*)

Date	Event
1922	Phantom Ranch built; Colorado River Compact signed by the states of the Colorado River basin.
1923	U.S. Geological survey Expedition, headed by Claude Birdseye and Emery Kolb, makes a scientific study of the river corridor with photographs by E. C. LaRue and L. R. Freeman.
1929	Navajo Bridge built.
1930s	Camelthorn invades the river corridor.
1932	Grand Canyon National Monument is expanded to include much of the western Grand Canyon.
1935–36	Hoover Dam is completed; Lake Mead National Recreation Area is created.
1938	Elzada Clover and Lois Jotter become the first biologists to float through the canyon, where they study vegetation on a research trip led by Norm Nevills.
1941	Lake Mead first reaches maximum pool elevation, backs water far up into Grand Canyon.
1942	The humpback chub is discovered as a species by Robert Rush Miller on the basis of a specimen from the mouth of Bright Angel Creek.
1953	Robert W. Dickerman becomes the first ornithologist to run the river and study its birdlife.
1956	Congress authorizes construction of Glen Canyon Dam with the Colorado River Storage Project Act.
late 1950s	Biological "salvage" studies are conducted in Glen Canyon by Angus M. Woodbury and others from the University of Utah.
1962	The U.S. Fish and Wildlife Service sterilizes 500 miles of the Green River in Utah with the poison rotenone, killing all aquatic life; indirect effects on the Colorado River in Grand Canyon are unknown but are probably detrimental to native fish and other aquatic life.
1963	River outlet tubes are closed at Glen Canyon Dam; Lake Powell begins to form.
1964	Glen Canyon Dam begins generation of electricity.
1965	U.S. Geological Survey Expedition with Luna Leopold studies the pools and rapids of the river.
1966	Crystal Creek Rapid greatly enlarged by a debris flow.
1968	Congress passes the Colorado River Basin Act.

Table B.1   (*Continued*)

Date	Event
1969	Marble Canyon National Monument created; Congress decides not to build Marble Canyon and Bridge Canyon (Hualapai) dams.
1969–70	Robert C. Euler directs excavations at Stanton's Cave with funding from the National Geographic Society.
1970	Bright Angel (silver) Bridge built across river at Phantom Ranch.
1973	The National Park Service freezes commercial and private river use at 1972 levels and initiates the Colorado River Research Program, directed by R. Roy Johnson.
1975	Grand Canyon National Park enlarged by Congress to include Marble Canyon, Toroweap area, and much of upper Lake Mead; the park now encompasses the entire river corridor from Lees Ferry to the Grand Wash Cliffs.
1980	Lake Powell reaches its designed maximum capacity; the dam releases 49,000 cfs.
1981	The Bureau of Reclamation Peaking Power proposal fails. Feral burros are removed from the park by the Fund for Animals.
early 1980s	Bald eagles begin overwintering at the mouth of Nankoweap Creek on an annual basis.
1982	Interior Secretary James Watt authorizes the Glen Canyon Environmental Studies program to study the short- and long-term effects of Glen Canyon Dam on the river environment in Grand Canyon.
1983	Glen Canyon Dam releases more than 92,000 cfs to stop Lake Powell from overtopping the dam; flood is the largest through the canyon since the dam was completed in 1963.
1984	One of the last razorback suckers seen in Grand Canyon is caught and released near Bass Rapid.
1988	The U.S. Fish and Wildlife Service formally determines that the operation of Glen Canyon Dam jeopardizes the continued existence of the humpback chub along the river in Grand Canyon.
1989	U.S. Federal District Court in Salt Lake City rules that the Western Area Power Administration cannot enter into long-term power contracts with power users without first implementing an Environmental Impact Statement.

Table B.1   (*Continued*)

Date	Event
1990	U.S. Bureau of Reclamation and Western Area Power Administration begin conducting Environmental Impact Statements on their operations as required by the National Environmental Policy Act.

SUGGESTED READINGS

Crumbo, K. *A River Runner's Guide to the History of the Grand Canyon.* Boulder, Colo.: Johnson Books, 1981.

Hughes, J. D. *In the House of Stone and Light: A Human History of the Grand Canyon.* Grand Canyon, Ariz.: Grand Canyon Natural History Association, 1978.

Lavender, D. *River Runners of the Grand Canyon.* Grand Canyon, Ariz.: Grand Canyon Natural History Association; Tucson: University of Arizona Press, 1985.

Stevens, L. *The Colorado River in Grand Canyon: A Guide.* Flagstaff, Ariz.: Red Lake Books, 1983.

NOTES

Introduction

1. For more information, see R. B. Stanton, "Through the Grand Canyon of the Colorado," *Scribners Monthly Magazine* 8 (1890): 591–613.

2. For a summary of research on Stanton's Cave, see R. C. Euler, ed., *The Archaeology, Geology, and Paleobiology of Stanton's Cave*, Monograph No. 6 (Grand Canyon, Ariz.: Grand Canyon Natural History Association, 1984).

3. That the climate and vegetation along the Colorado River of ten thousand years ago has changed has been thoroughly documented. See P. S. Martin "Stanton's Cave During and After the Last Ice Age," in Euler, *Archaeology, Geology, and Paleobiology of Stanton's Cave*, pp. 131–37.

4. For more information on the recreational use of the river, see P. L. Fradkin, *A River No More: The Colorado River and the West* (Tucson: University of Arizona Press, 1984).

5. See R. R. Johnson, *Synthesis and Management Implications of the Colorado River Research Program*, Colorado River Technical Report No. 17 (Grand Canyon, Ariz.: Grand Canyon National Park, 1977).

6. A concise summary of these changes through the early 1980s may be found in S. W. Carothers and R. Dolan, "Dam Changes on the Colorado River," *Natural History Magazine* 91 (1982)(1): 74–83.

7. For a review of the Glen Canyon Environmental Studies, see National Academy of Sciences, *River and Dam Management: A Review of the Bureau of Reclamation's Glen Canyon Environmental Studies* (Washington, D.C.: National Academy Press, 1987).

8. Fradkin, *A River No More*, p. 15.

Chapter 1. The River: Dramatic Rise and Fall

1. L. R. Freeman, "Surveying the Grand Canyon of the Colorado: An Account of the 1923 Boating Expedition of the United States Geological Survey," *National Geographic Magazine* 45 (1924) (5): 524–25.

2. For purposes of comparison, a small stream 3 feet across, 1

foot deep, and flowing 1 foot per second would be running 3 cubic feet per second (3 times 1 times 1).

3. Freeman, "Surveying the Grand Canyon," p. 526.

4. J. W. Powell, *Exploration of the Colorado River of the West and Its Tributaries* (Washington, D.C.: U.S. Government Printing Office, 1875), p. 62.

5. This theory of the control and maintenance of Grand Canyon rapids has now been disproved, but for more information see L. B. Leopold, "The Rapids and the Pools — Grand Canyon," in *The Colorado River Region and John Wesley Powell*, U.S. Geological Survey Professional Paper 668, 1969, pp. 131–45.

6. R. Dolan, A. Howard, and D. Trimble, "Structural Control of the Rapids and Pools of the Colorado River in Grand Canyon," *Science* 202 (1978): 629–31.

7. Steve Carothers witnessed the creation of such a rapid in March 1974 when a rock fall just below Tiger Wash (River Mile 27) exploded from the canyon wall. With a spectacular splash, the debris collapsed into the river, leaving the channel littered with car-sized boulders. The flow of the river at that time was low, about 5,000 cfs or lower, making for an exciting and difficult first run through the new rapid.

8. S. W. Kieffer, *The Rapids and Waves of the Colorado River, Grand Canyon, Arizona*, U.S. Geological Survey Open-File Report 87-096, 1987.

9. D. L. Smith and C. G. Crampton, eds., *The Colorado River Survey: Robert B. Stanton and the Denver, Colorado Canyon, and Pacific Railroad* (Salt Lake City: Howe Brothers, 1987), p. 81.

10. For more information, see Kieffer, *The Rapids and Waves of the Colorado River.*

11. J. W. Powell, *The Exploration of the Colorado River and Its Canyons* (New York: Penguin Books, 1987), p. 218.

12. W. K. Hamblin, of Brigham Young University, provided this information, based on his forthcoming book: *Late Cenozoic Lava Dams in the Western Grand Canyon*, Brigham Young University Geology Studies, Special Publication No. 8.

Chapter 2. Sand and Rock: Sediment and Substrate

1. The December 1966 storm also caused debris flows in Bright Angel Canyon and other canyons draining the Kaibab Plateau. See M. E. Cooley, B. N. Aldridge, and R. C. Euler, *Effects of the Catastrophic Flood of December, 1966, North Rim Area, Eastern Grand Canyon, Arizona*, U.S. Geological Survey Professional Paper 980, 1977.

2. A reassessment of the effects of the December 1966 storm was made in R. H. Webb, P. T. Pringle, and G. R. Rink, *Debris Flows*

from Tributaries of the Colorado River, Grand Canyon National Park, Arizona, U.S. Geological Survey Open-File Report 87-118, 1987.

3. Only rapids rated at five or greater (ratings from L. Stevens, *The Colorado River in Grand Canyon: A Guide* [1983]) were considered. See R. H. Webb, P. T. Pringle, S. L. Reneau, and G. R. Rink, "Monument Creek Debris Flow, 1984: Implications for Formation of Rapids on the Colorado River in Grand Canyon National Park," *Geology* 16 (1988): 50–54.

4. Webb et al., *Debris Flows*.

5. Ibid.

6. J. C. Schmidt and J. B. Graf, *Aggradation and Degradation of Alluvial Sand Deposits, 1965 to 1986, Colorado River, Grand Canyon National Park, Arizona*, U.S. Geological Survey Open-File Report 87-555, 1988.

7. For more information, see W. O. Smith et al., *Comprehensive Survey of Sedimentation in Lake Mead, 1948–49*, U.S. Geological Survey Professional Paper 295, 1960.

8. A. Howard and R. Dolan, "Geomorphology of the Colorado River in the Grand Canyon," *Journal of Geology* 89 (1981): 269–98.

9. Webb et al., *Debris Flows*.

10. Kieffer, *Rapids and Waves of the Colorado River*.

11. This scenario was devised from changes recorded at the Lees Ferry gauging station. See D. E. Burkham, *Trends in Selected Hydraulic Variables for the Colorado River at Lees Ferry and Near Grand Canyon for the Period 1922–1984*, Glen Canyon Environmental Studies Technical Report (Salt Lake City: Bureau of Reclamation, 1986).

12. A streambed is referred to as *armored* when the fine sediments such as silt and sand have been eroded away, leaving only the larger cobbles and boulders. At this point the streambed is virtually resistant to further erosion, hence the term *armored*.

13. Howard and Dolan, "Geomorphology of the Colorado River."

14. The existing information on beach erosion is extensive and occasionally contradictory. See S. S. Beus, S. W. Carothers, and C. C. Avery, "Topographic Changes in Fluvial Terrace Deposits Used in Campsite Beaches Along the Colorado River in Grand Canyon," *Journal of the Arizona-Nevada Academy of Science* 20 (1985): 111–20; and Schmidt and Graf, *Aggradation and Degradation of Alluvial Sand Deposits*.

15. This estimate was generated by a computer model of sediment transport in the river. See T. J. Randle and E. L. Pemberton, *Results and Analysis of the STARS Modeling Efforts of the Colorado River in Grand Canyon*, Glen Canyon Environmental Studies Technical Report (Salt Lake City: Bureau of Reclamation, 1987).

16. A. D. Howard, *Establishment of Bench Mark Study Sites Along the Colorado River in Grand Canyon National Park for Monitoring of Beach Erosion Caused by Natural Forces and Human Impact,* University of Virginia Grand Canyon Study, Technical Report No. 1, 1975.

17. Beus et al., "Topographic Changes in Fluvial Terrace Deposits."

18. This conclusion springs from a study of beach composition begun in the 1970s: L. E. Stevens and G. L. Waring, *Effects of Post-Dam Flooding on Riparian Substrate, Vegetation, and Invertebrate Populations in the Colorado River Corridor in Grand Canyon, Arizona,* Glen Canyon Environmental Studies Technical Report (Salt Lake City: Bureau of Reclamation, 1986).

19. There is no scientific evidence that the beaches will disappear in any definite period of time. For an estimate, see E. M. Laursen and E. Silverston, *Hydrology and Sedimentology of the Colorado River in Grand Canyon,* Colorado River Technical Report No. 13 (Grand Canyon, Ariz.: Grand Canyon National Park, 1976).

Chapter 3. Building Blocks: Primary Producers and Aquatic Invertebrates

1. D. M. Kubly and G. A. Cole, "The Chemistry of the Colorado River and Its Tributaries in Marble and Grand Canyons," *Proceedings of the First Conference on Scientific Research in the National Parks* (New Orleans, November 9–12, 1976), U.S. Department of the Interior, National Park Service Transactions and Proceedings Series 1 (1979) (5): 565–72.

2. J. V. Ward and J. A. Stanford, *The Ecology of Regulated Streams* (New York: Plenum Press, 1979).

3. See E. M. Laursen, S. Ince, and J. Pollack, "On Sediment Transport Through the Grand Canyon," *Proceedings of the Third Federal Interagency Sedimentation Conference* (Denver, March 1976), 1976; E. M. Laursen and E. Silverston, *Hydrology and Sedimentation of the Colorado River in Grand Canyon,* Technical Report No. 13 (Grand Canyon, Ariz.: Grand Canyon National Park, 1976); E. L. Pemberton, "Channel Changes in the Colorado River Below Glen Canyon," *Proceedings of the Third Federal Interagency Sedimentation Conference* (Denver, March 1976), 1976.

4. W. V. Iorns, C. H. Hembree, and G. L. Oakland, *Water Resources of the Upper Colorado River Basin — Technical Report,* U.S. Geological Survey Professional Paper 441, 1965.

5. L. J. Paulson and T. D. Evans, "The Influence of Lake Powell on the Suspended Sediment-Phosphorus Dynamics of the Colorado River Inflow to Lake Mead," in D. V. Adams and V. A. Lamarra, eds., *Aquatic Resources Management of the Colorado River Ecosystem: Proceedings of the 1981 Symposium on the Aquatic Resources*

Management of the Colorado River Ecosystem (November 16–18, 1981, Las Vegas, Nev.), pp. 57–70, Office of Water and Research Technology, Utah Water Research Laboratory and Utah State University, 1983.

6. H.B.N. Hynes, *The Ecology of Running Waters* (Toronto: University of Toronto Press, 1970).

7. S. Flowers, "List of Algae Collected from Glen Canyon," in *Ecological Studies of the Flora and Fauna in Glen Canyon* (Salt Lake City: University of Utah Anthropology Papers, 1959).

8. L. G. Williams, *Plankton Population Dynamics*, Public Health Publication 663, Suppl. 2, Washington, D.C., 1962.

9. M. R. Sommerfeld, W. M. Crayton, and N. L. Crane, *Survey of Bacteria, Phytoplankton and Trace Chemistry of the Lower Colorado River and Tributaries in the Grand Canyon National Park*, Technical Report No. 12 (Grand Canyon, Ariz.: Grand Canyon National Park, 1976).

10. L. Haury, *Zooplankton of the Colorado River: Glen Canyon Dam to Diamond Creek*, Glen Canyon Environmental Studies Technical Report (Salt Lake City: Bureau of Reclamation, 1987).

11. H. R. Maddux, D. M. Kubly, J. C. deVos, Jr., W. R. Persons, R. Staedicke, and R. L. Wright, *Effects of Varied Flow Regimes on Aquatic Resources of Glen and Grand Canyons*, Final Report by the Arizona Game and Fish Department to the Glen Canyon Environmental Studies (Salt Lake City: Bureau of Reclamation, 1987).

12. H. D. Usher, D. W. Blinn, G. G. Hardwick, and W. C. Leibfried, *Cladophora glomerata and Its Diatom Epiphytes in the Colorado River Through Glen and Grand Canyons: Distribution and Dessication Tolerance*, Glen Canyon Environmental Studies Technical Report (Salt Lake City: Bureau of Reclamation, 1987).

13. D. W. Blinn, R. Truitt, and A. Pickart, "Response of Epiphytic Diatom Communities from the Tailwaters of Glen Canyon Dam, Arizona, to Elevated Water Temperature," *Regulated Rivers: Research and Management* 4 (1989): 91–96.

14. D. B. Czarnecki, D. W. Blinn, and T. Tompkins, *A Periphytic Microflora Analysis of the Colorado River and Major Tributaries in Grand Canyon and Vicinity*, Colorado River Technical Report No. 6 (Grand Canyon, Ariz.: Grand Canyon National Park, 1976).

15. G. W. Hofknecht, "Seasonal Community Dynamics of Aquatic Invertebrates in the Colorado River and Its Tributaries Within Grand Canyon, Arizona" (M.S. thesis, Northern Arizona University, Flagstaff, 1981).

16. D. W. Blinn, C. Pinney, R. Truitt, and A. Pickart, *The Influence of Elevated Water Temperatures on Epiphytic Diatom Species in the Tailwaters of Glen Canyon Dam and the Importance of These Epiphytic Diatoms in the Diet of Gammarus lacustris*, Final Report (Salt Lake City: Bureau of Reclamation, 1986).

17. S. W. Carothers and C. O. Minckley, *A Survey of the Aquatic Flora and Fauna of the Grand Canyon*, Final Report to Water and Power Resources Service, Lower Colorado Region, Boulder City, Nev., 1981.

18. H. D. Usher and D. W. Blinn, "Influence of Various Exposure Periods on the Biomass and Chlorophyll *A* of *Cladophora glomerata* (Chlorophyta)," *Journal of Phycology* (in press).

19. Kubly and Cole, "Chemistry of the Colorado River and Its Tributaries."

20. Carbonates have also been suggested as the cause of the color of the Little Colorado River (Dean W. Blinn, Northern Arizona University, personal communication).

Chapter 4. Ecology of Aquatic Vertebrates

1. C. D. Bancroft and K. Sylvester, *The Colorado River Glen Canyon Tailwater Fishery*, Annual Report, Arizona Game and Fish Department, Region II, 1978.

2. The fish we know today as the Colorado River squawfish was referred to as the Colorado salmon by Kolb. See E. C. Kolb *Through the Grand Canyon from Wyoming to Mexico* (New York: Macmillan, 1914), p. 15.

3. Carothers and Minckley, *Survey of the Aquatic Flora and Fauna of the Grand Canyon.*

4. Maddux et al., *Effects of Varied Flow Regimes on Aquatic Resources of Glen and Grand Canyons.*

5. R. D. Suttkus and G. H. Clemmer, "The Fishes of the Colorado River in Grand Canyon National Park," *Proceedings of the First Conference on Scientific Research in the National Parks* (New Orleans, November 9–12, 1976), U.S. Department of the Interior, National Park Service Transactions and Proceedings Series 1 (1979) (5): 599–604.

6. R. R. Williamson and C. F. Tyler, "Trout Propagation in Grand Canyon National Park," *Grand Canyon Nature Notes* 7 (1932) (2): 11–16.

7. S. W. Carothers, "Too Thick to Drink, Too Thin to Plow," *Arizona Wildlife*, Arizona Game and Fish Department, pp. 4–10, 1985.

8. R. R. Miller, "Origin and Affinities of the Freshwater Fish Fauna of Western North America," in C. L. Hubbs, ed., *Zoogeography, American Association for the Advancement of Science Publications* 51 (1958): 187–222; R. R. Miller, "Man and the Changing Fish Fauna of the American Southwest," *Papers of the Michigan Academy of Science, Arts, and Letters* 46 (1961): 365–404.

9. M. Molles, "The Impacts of Habitat Alterations and Introduced Species on the Native Fishes of the Upper Colorado

River Basin," in W. D. Spofford, Jr., A. L. Parker, and A. V. Kneese, eds., *Energy Development in the Southwest, Volume 2*, Resources for the Future, Resource Paper R-18, pp. 163–81 (Baltimore: John Hopkins University, 1980).

10. Ellsworth Kolb and Emery Kolb, "Experiences in the Grand Canyon," *National Geographic Magazine* 26 (1914)(2): 127.

11. R. D. Suttkus and G. H. Clemmer, "The Humpback Chub, *Gila cypha*, in the Grand Canyon Area of the Colorado River," *Occasional Papers of Tulane University, Museum of Natural History* 1 (1977): 1–30; L. R. Kaeding and M. A. Zimmerman, "Life History and Ecology of the Humpback Chub in the Little Colorado and Colorado Rivers of the Grand Canyon," *Transactions of the American Fisheries Society* 112 (1983): 577–94.

12. R. J. Behnke and D. E. Benson, *Endangered and Threatened Fishes of the Upper Colorado River Basin*, Bulletin 503A, Cooperative Extension Service (Fort Collins Colorado State University, 1980).

13. P. B. Holden and C. B. Stalnaker, "Systematic Studies of the Cyprinid Genus *Gila*, in the Upper Colorado River Basin," *Copeia* 1970(3): 409–20; P. B. Holden, "A Study of the Habitat Use and Movement of the Rare Fishes in the Green River, Utah," *Transactions, Bonneville Chapter of the American Fisheries Society* 1978: 64–89; R. A. Valdez, *Cataract Canyon Fish Study*, Final Report (Salt Lake City: Bureau of Reclamation, 1985).

14. Miller, "Man and the Changing Fish Fauna."

15. W. L. Minckley, 1985. *Native Fishes and Natural Aquatic Habitats in U.S. Fish and Wildlife Service Region II, West of the Continental Divide*, Internal Report (Albuquerque: U.S. Fish and Wildlife Service, Division of Endangered Species, 1985).

16. Kolb, *Through the Grand Canyon*, p. 15.

17. R. R. Miller, "Is Our Native Underwater Life Worth Saving?" *National Parks Magazine* 37 (1963)(188): 5.

18. K. B. Clifton, "The Route of James O. Pattie on the Colorado in 1826," *Arizona and the West* 6 (1964): 119–36.

19. D. Lavender, *River Runners of the Grand Canyon* (Grand Canyon, Ariz.: Grand Canyon Natural History Association; Tucson: University of Arizona Press, 1985).

20. G. F. Flavell, *The Log of the Pantheon: An Account of an 1896 River Voyage from Green River, Wyoming, to Yuma, Arizona, Through the Grand Canyon*, ed. N. B. Carmony and D. E. Brown (Boulder, Colo.: Pruett Publishing, 1987).

21. Lavender, *River Runners of the Grand Canyon*.

22. Freeman, "Surveying the Grand Canyon of the Colorado."

23. Otis ("Doc") Marston discovered the identity of the trappers who left their outfit in the cave. Frederik Tyler Barry and two others arrived by boat but later abandoned their gear at this site in 1888. They left on foot, evidently on the left (south) bank, and later

reached Flagstaff, Arizona. Information source: National Park Service archives, Grand Canyon Village.

24. Robert Euler encountered a party of three men, armed with rifles and dressed in buckskin, on horseback at the mouth of Parashant Canyon in June 1960. They indicated that they were living off the land, which probably included trapping.

25. V. O. Bailey, *Mammals of the Grand Canyon Region*, Bulletin No. 1 (Grand Canyon, Ariz.: Grand Canyon Natural History Association, 1935).

26. G. A. Ruffner, N. J. Czaplewski, and S. W. Carothers, "Distribution and Natural History of Some Mammals from the Inner Gorge of the Grand Canyon, Arizona," *Journal of the Arizona-Nevada Academy of Science* 13 (1978): 85–91.

27. D. F. Hoffmeister, *Mammals of Arizona* (Tucson: University of Arizona Press; Arizona Game and Fish Department, 1986).

28. S. W. Carothers, personal observations.

29. Euler, *Archaeology, Geology, and Paleobiology of Stanton's Cave.*

30. Hoffmeister, *Mammals of Arizona.* Hoffmeister mentions that Nelson visited Lees Ferry in July 1909, when he saw one otter and some tracks. Nelson's original field journal in the Smithsonian Institution Archives placed him at Lees Ferry from August 22–27, 1909. His journal reads: "Now and then a beaver, otter or muskrat occurs along the river." This mention of muskrat is the only sighting of the species until a specimen was collected in 1986.

31. Kolb, *Through the Grand Canyon from Wyoming to Mexico*, p. 259.

Chapter 5. Green on Red: Riparian Plants in a Desert Canyon

1. Bessie Hyde was the first woman to float the Colorado River in Grand Canyon, in 1928. Her trip ended tragically, however, probably at 232 Mile Rapid, and she did not complete the full voyage through the canyon. See K. Crumbo, *A River Runner's Guide to the History of the Grand Canyon* (Boulder, Colo.: Johnson Books, 1981).

2. For more details on the expedition, see Lavender, *River Runners of the Grand Canyon.*

3. E. U. Clover and L. Jotter, "Cacti of the Colorado River and Tributaries," *Bulletin of the Torrey Botanical Club* 68 (1941): 409–19; E. U. Clover and L. Jotter, "Floristic Studies in the Canyon of the Colorado and Its Tributaries," *American Midland Naturalist* 32 (1944): 591–642.

4. Clover and Jotter, "Floristic Studies in the Canyon of the Colorado," p. 620.

5. R. R. Johnson and D. A. Jones, technical coordinators, *Importance, Preservation, and Management of Riparian Habitat: A*

Symposium, USDA Forest Service General Technical Report RM-43, Fort Collins, Colo., 1977.

6. B. G. Phillips, A. M. Phillips III, and M.A.S. Bernzott, *Annotated Checklist of Vascular Plants of Grand Canyon National Park*, Monograph No. 7 (Grand Canyon, Ariz.: Grand Canyon Natural History Association, 1987).

7. Clover and Jotter, "Floristic Studies in the Canyon of the Colorado," p. 615.

8. For more details of vegetation changes in the Stanton's Cave area, see Euler, *Archaeology, Geology, and Paleobiology of Stanton's Cave*. For related information with a broader application to the entire Grand Canyon, see: K. L. Cole, "Late Quaternary Environments in the Eastern Grand Canyon: Vegetational Gradients over the Last 25,000 Years" (Ph.D. diss., University of Arizona, 1981).

9. Scouring floods effectively removed marshes on an annual basis before the dam. J. C. Schmidt, B. T. Brown, and L. E. Stevens, *Regulated Discharge and Marsh Development Along the Colorado River in Grand Canyon, Arizona* (in preparation).

10. These trends are from selected study sites and may not accurately reflect changes in the river corridor as a whole. See M. J. Pucherelli, *Evaluation of Riparian Vegetative Trends in the Grand Canyon Using Multitemporal Remote Sensing Techniques*, Glen Canyon Environmental Studies Technical Report (Salt Lake City: Bureau of Reclamation, 1986).

11. R. M. Turner and M. M. Karpiscak, *Recent Vegetation Changes Along the Colorado River Between Glen Canyon Dam and Lake Mead, Arizona*, U.S. Geological Survey Professional Paper 1132, 1980.

12. W. L. Graf, "Fluvial Adjustments to the Spread of *Tamarisk* in the Colorado Plateau Region," *Geological Society of America Bulletin* 87 (1978): 1491–1501.

13. Turner and Karpiscak, *Recent Vegetation Changes Along the Colorado River*.

14. Most ecological information on coyote willow before the flood of 1983 came from N. J. Brian, "A Preliminary Study of the Riparian Coyote Willow Communities Along the Colorado River in Grand Canyon National Park, Arizona" (M.S. thesis, Northern Arizona University, Flagstaff, 1982).

15. Estimates of the amount of vegetation lost during the 1983 flood vary, depending on the methods used. See Pucherelli, *Evaluation of Riparian Vegetative Trends in the Grand Canyon*; Brian, "Riparian Coyote Willow Communities Along the Colorado River."

16. Stevens and Waring, *Effects of Post-Dam Flooding*.

17. Tamarisk is not as tolerant of submersion by warm water as it is by cold. Along the margins of upper Lake Mead in 1973–74 when the lake was high for a few months, inundating many acres of tamarisk, there were few survivors. In Grand Canyon the cold

waters of the Colorado River apparently induced a degree of physiological dormancy that left a relatively high rate of survivorship when the floodwater receded (Arthur M. Phillips, III, personal communication).

18. Information on seed germination and seedling establishment has been summarized from Stevens and Waring, *Effects of Post-Dam Flooding.*

19. Recent studies addressing this question have been inconclusive. See L. S. Anderson and G. A. Ruffner, *Effects of the Post–Glen Canyon Dam Flow Regime on the Old High-Water Line Plant Community Along the Colorado River in Grand Canyon,* Glen Canyon Environmental Studies Technical Report, (Salt Lake City: Bureau of Reclamation, 1987).

20. Much of the decline of the old high-water zone took place upstream of Phantom Ranch at the edge of the range of mesquite and acacia, where freezing winter temperatures may have played a role in the total loss of foliage. See also Pucherelli, *Evaluation of Riparian Vegetative Trends in the Grand Canyon.*

Chapter 6. Riparian Insects: Ants, Black Flies, Leafhoppers, and More

1. J. O. Schmidt and M. F. Blum, "A Harvester Ant Venom: Chemistry and Pharmacology," *Science* 200 (1978): 1064–66.

2. B. P. Hayden, P. Dolan, and S. Carothers, "Float-Trip Campsites, Red Harvester Ants, and the Common Ant Lion: Man's Impact on Food Chains," *Grand Canyon Studies,* Museum of Northern Arizona Manuscript Report, pp. 16–25, 1977.

3. S. W. Carothers and S. W. Aitchison, *An Ecological Survey of the Riparian Zone of the Colorado River Between Lees Ferry and the Grand Wash Cliffs, Arizona,* Colorado River Technical Report No. 10 (Grand Canyon, Ariz.: Grand Canyon National Park, 1976).

4. U.S. Department of the Interior, National Park Service, *Draft Environmental Statement, Proposed Colorado River Management Plan* (Grand Canyon, Ariz.: Grand Canyon National Park, 1977); U.S. Department of the Interior, National Park Service, *Colorado River Management Plan* (Grand Canyon, Ariz.: Grand Canyon National Park, 1979).

5. S. S. Beus et al., *Colorado River Investigations I–VIII,* Northern Arizona University, Department of Geology, 1982–89. (This series of unpublished reports is available at the Museum of Northern Arizona, Flagstaff.)

6. W. M. Wheeler, *Ants: Their Structure, Development, and Behavior,* Columbia University Biological Series, IX (New York: Columbia University Press, 1926). There is always the exception to the rule: many wasps are both colonial and carnivorous, and some other predatory insects defend territories.

7. C. Pike, R. LaChat, and C. O'Rourke Taylor, "Continued Studies on the Red Harvester Ant: Density and Foraging Activities on Human Impacted, Colorado River Beaches in Grand Canyon National Park, Summer 1989," in S. S. Beus, L. E. Stevens, and F. B. Lojko, *Colorado River Investigations VIII: July–August, 1989,* Northern Arizona University, 1989.

8. That is the hypothesis of Larry Stevens, an ecologist who has studied the insects of the river corridor.

9. L. E. Stevens, "An Insect Inventory of Grand Canyon" and "Insect Production on Native and Introduced Dominant Plant Species," in Carothers and Aitchison, *Ecological Survey of the Riparian Zone of the Colorado River;* L. E. Stevens "Invertebrate Herbivore Community Dynamics on *Tamarix chinensis* Loueiro and *Salix exigua* Nuttal in the Grand Canyon, Arizona" (M.S. thesis, Northern Arizona University, Flagstaff, 1985).

10. Susan C. Jones, personal communication. See S. C. Jones, "New Termite Records for the Grand Canyon," *Southwestern Entomologist* 10 (1985): 137–38.

11. Larry Stevens, personal communication.

12. Stevens, "Invertebrate Herbivore Community Dynamics."

13. Larry Stevens, personal communication.

14. Many large-bodied riparian-nesting birds forage heavily on cicadas, and some species of birds even time their nesting cycle to coincide with cicada emergence. This behavior has not yet been documented along the Colorado River in Grand Canyon, but it is very likely. See R. L. Glinski and R. D. Ohmart, "Breeding Ecology of the Mississippi Kite in Arizona," *Condor* 85 (1983): 200–207.

15. R. L. Glinski and R. D. Ohmart, "Factors of Reproduction and Population Densities in the Apache Cicada (*Diceroprocta apache*)," *Southwestern Naturalist* 29 (1984) (1): 73–79.

Chapter 7. Wildlife of the River Corridor

1. For more information, see B. T. Brown, R. Mesta, L. E. Stevens, and J. Weisheit, "Changes in the Winter Distribution of Bald Eagles Along the Colorado River in Grand Canyon, Arizona," *Journal of Raptor Research* 23 (1989): 110–13.

2. Although the increase in wintering eagles at Nankoweap is analogous to the increase in Glacier National Park, the Arizona concentration is unlikely ever to become as large as the one in Montana. See B. R. McClelland, "Autumn Concentrations of Bald Eagles in Glacier National Park," *Condor* 75 (1973): 121–23.

3. B. T. Brown, S. W. Carothers, and R. R. Johnson, *Grand Canyon Birds: Historical Notes, Natural History, and Ecology* (Tucson: University of Arizona Press, 1987).

4. B. T. Brown, S. W. Carothers, and R. R. Johnson, "Breeding

Range Expansion of Bell's Vireo in Grand Canyon, Arizona," *Condor* 85 (1983): 499–500.

5. The willow flycatcher was being considered as a federally endangered species by the U.S. Fish and Wildlife Service in 1990. For more information on its nesting ecology in the river corridor, see B. T. Brown, "Breeding Ecology of a Willow Flycatcher Population Along the Colorado River in Grand Canyon, Arizona," *Western Birds* 19 (1988): 25–33.

6. For a complete discussion of the nesting birds of the river corridor and their abundance in different habitats, see B. T. Brown and R. R. Johnson, *Fluctuating Flows from Glen Canyon Dam and Their Effect on Breeding Birds of the Colorado River in Grand Canyon, Arizona*, Glen Canyon Environmental Studies Technical Report (Salt Lake City: Bureau of Reclamation, 1987).

7. D. F. Hoffmeister, *Mammals of Grand Canyon* (Urbana: University of Illinois Press, 1971).

8. G. A. Ruffner, N. J. Czaplewski, and S. W. Carothers, "Distribution and Natural History of Some Mammals from the Inner Gorge of the Grand Canyon, Arizona," *Journal of the Arizona-Nevada Academy of Science* 13 (1978): 85–91.

9. P. L. Warren and C. R. Schwalbe, *Lizards Along the Colorado River in Grand Canyon National Park: Possible Effects of Fluctuating River Flows*, Glen Canyon Environmental Studies Technical Report (Salt Lake City: Bureau of Reclamation, 1986).

10. Ibid.

11. The percentages of the various bird populations that were lost varied with the timing of the nesting season, as well as the height of nests above ground. See B. T. Brown and R. R. Johnson, "Glen Canyon Dam, Fluctuating Water Levels, and Riparian Breeding Birds: The Need for Management Compromise on the Colorado River in Grand Canyon," *Proceedings of the First North American Riparian Conference*, USDA Forest Service General Technical Report RM-120 (1985): 76–80.

12. B. T. Brown, *Monitoring Bird Population Densities Along the Colorado River in Grand Canyon: 1987 Breeding Season*, Glen Canyon Environmental Studies Technical Report (Salt Lake City: Bureau of Reclamation, 1988).

13. Warren and Schwalbe, *Lizards Along the Colorado River*.

14. See Euler, *Archaeology, Geology, and Paleobiology of Stanton's Cave*.

15. J. I. Mead, "The Last 30,000 Years of Faunal History Within the Grand Canyon, Arizona," *Quaternary Research* 15 (1981): 311–26; J. I. Mead and A. M. Phillips III, "The Late Pleistocene and Holocene Fauna of Vulture Cave, Arizona," *Southwestern Naturalist* 26 (1981): 257–88; S. D. Emslie, "Age and Diet of Fossil California Condors in Grand Canyon, Arizona," *Science* 237 (1987): 768–70.

16. The effect of feral burros on desert environments is a subject of sometimes bitter dispute. See S. W. Carothers, M. E. Stitt, and R. R. Johnson, "Feral Asses on Public Lands: An Analysis of the Biotic Impact, Legal Considerations, and Management Alternatives," *Transactions of the North American Wildlife and Natural Resources Conference* 41 (1976): 396–406.

17. Hoffmeister, *Mammals of Arizona.*

18. D. M. Miller, R. A. Young, T. W. Gatlin, and J. A. Richardson, *Amphibians and Reptiles of the Grand Canyon*, Monograph No. 4 (Grand Canyon, Ariz.: Grand Canyon Natural History Association, 1982).

19. Speculation on the effects of the river as a barrier to wildlife began in the 1930s. See Hoffmeister, *Mammals of Arizona*; E. A. Goldman, "The Colorado River as a Barrier to Mammalian Distribution," *Journal of Mammalogy* 18 (1937): 427–35.

20. The river as a barrier to subspecies is not an all-or-nothing rule. Generally, populations occurring on opposite sides are distinct, but individuals resembling the population on the opposite side of the river may exist as exceptions within a few miles of the river. Information on subspecies separated by the river is from range maps in Hoffmeister, *Mammals of Arizona.*

21. Fifty-eight pairs of peregrine falcons were found nesting in 1989 in an area comprising only 25 percent of the park. Thirty-seven of them were in the river corridor. See B. T. Brown, *Grand Canyon Peregrine Falcon Population Study: 1989 Interim Report*, Final Report (Grand Canyon, Ariz.: Grand Canyon National Park, 1989).

Chapter 8. Biopolitical Management of the River

1. For more information, see N. Hundley, Jr., *Water and the West: The Colorado River Compact and the Politics of Water in the American West* (Berkeley: University of California Press, 1975).

2. The upper and lower basin states split the amount due to Mexico, with the upper basin states deducting the amount of water provided by the Paria River.

3. The Western Area Power Administration, in its marketing of hydropower from Glen Canyon Dam during the late 1970s and 1980s, may have misinterpreted the intent of Congress with respect to the purposes of the dam. The Colorado River Basin Project Act of 1968, 43 U.S.C. Sections 1501 et seq., states:

> This program is declared to be for the purposes, among others, of regulating the flow of the Colorado River; controlling floods; improving navigation; providing for the storage and delivery of waters of the Colorado River for reclamation of lands, including supplemental water supplies, and for municipal, industrial, and other beneficial purposes; improving water quality; *providing for*

basic public outdoor recreation facilities; improving conditions for fish and wildlife, and the generation and sale of electrical power as an incident of the foregoing purposes. [Emphasis added.]

Although the provisions of the act of 1968 are somewhat at odds with those of the act of 1956, the Department of the Interior's regional solicitor in Salt Lake City, in a memorandum dated June 6, 1988, to the director of the Bureau of Reclamation for the Upper Colorado Region, pointed out that the 1968 act specifically provides that the 1956 act shall be administered in accordance with the criteria established by the 1968 act.

4. U.S. Department of the Interior, *Glen Canyon Environmental Studies: Final Report* (Salt Lake City: Bureau of Reclamation, 1988), pp. D-1–D-49.

5. Information on the history, purpose, revenues, and scope of WAPA has been obtained from Western Area Power Administration, *1988 Annual Report* (Golden, Colo.: Western Area Power Administration, 1988).

6. The National Park Service expressed no opposition to the construction of Glen Canyon Dam or any of the three large dams proposed for the Grand Canyon region in the 1940s and 1950s. Its only concern was that the presence of dams and reservoirs might have a negative impact on the scenery. Frederick Law Olmsted, a National Park Service consultant hired to assess this consideration, concluded that the dams would not mar the scenery of the park itself and might even help visitors to get a better view of the Grand Canyon. See U.S. Department of the Interior, *The Colorado River: A Natural Menace Becomes a National Resource*, Departmental Report, U.S. Bureau of Reclamation (Washington, D.C.: U.S. Government Printing Office, 1946), pp. 241–42.

7. Resource preservation is the priority of the National Park Service by mandate, but recreation appears to have the real priority in the management of the river by the National Park Service.

8. An Environmental Assessment on the uprate and rewind of the generators at Glen Canyon Dam was completed in December 1982, as required by the National Environmental Policy Act, resulting in a finding of "no significant impact." For a concise history of this process, see U.S. Department of the Interior, *Glen Canyon Environmental Studies: Final Report*, p. D-24.

9. This information was provided by the Western Area Power Administration to reflect the scope of its operations in the late 1980s.

10. Power from Glen Canyon Dam is sold based on long-term contract prices, spot market prices, and emergency prices.

11. This ruling came in response to a lawsuit filed in December 1988 against WAPA by the Grand Canyon Trust, the Western River Guides Association, the National Wildlife Federation, and American Rivers.

12. The assessment of environmental impacts may come in the form of a full Environmental Impact Statement, an Environmental Assessment, or other action.

Epilogue

1. J. D. Hughes, *In the House of Stone and Light: A Human History of the Grand Canyon* (Grand Canyon, Ariz.: Grand Canyon Natural History Association, 1978), p. 28.

2. J. C. Ives, *Report upon the Colorado River of the West; Explored in 1857 and 1858* (Washington, D.C.: U.S. Government Printing Office, 1861), p. 110.

3. That "National Parks should represent a vignette of primitive America" was the key theme of the Leopold Commission. See S. A. Cain, C. M. Cottam, I. N. Gabrielson, T. L. Kimball, and A. S. Leopold (chairman), *Wildlife Management in the National Parks*, unpublished report submitted to the Secretary of the Interior, Washington, D.C., 1963.

4. A successor of sorts to the Leopold Report of 1963, the National Parks and Conservation Association fulfilled National Park Service Director William Penn Mott's request for a contemporary assessment of the future role of research and resource management in the national parks. The result was a convincing case for holistic ecosystem management in the parks. See Commission on Research and Resource Management Policy in the National Park System, *National Parks: From Vignettes to a Global View* (Washington, D.C.: National Parks and Conservation Association, 1989).

5. The Northwest Power Planning Council is based in Portland, Oregon. Information on the 1980 act was obtained from the council's 1989 annual report.

INDEX

ABOUT THE AUTHORS

Steven W. Carothers has been directly involved with many aspects of ecological research and resource management of the Colorado River through Grand Canyon since the late 1960s. He is the founder and president of SWCA Environmental Consultants, Inc., and has been head of the biology department at the Museum of Northern Arizona, research scientist with the National Park Service, adjunct professor of biology at Northern Arizona University, and adjunct professor of natural resources at the University of Arizona. He received a Ph.D. in zoology from the University of Illinois in 1974.

Bryan T. Brown, a consulting research biologist, has studied the Colorado River through Grand Canyon, primarily its birds and their habitats, since 1976. From 1976 to 1987 he worked as an ecologist with the National Park Service. He received a Ph.D. in wildlife ecology from the University of Arizona in 1987. Together, Carothers and Brown have directed or participated in a combined total of more than one hundred raft trips through the Grand Canyon to study the river corridor.